Praise for *Boys & Sex*

"Eye-opening. . . . Every few pa ttle bit. . . . Even in the most anx *Sex*, it's clear that Orenstein believe nen they can become, and she believes in us, as parents, to raise them. As for how? There's a great, galvanizing chapter at the end."

—*New York Times Book Review*

"A brisk, bracing read, *Boys & Sex* is essential for any parent who wants their son to avoid the traps of what's become known as 'toxic masculinity,' to discover his true sexual orientation and desires without trauma, and to have not just safe, legal sex but really great sex. The whip-smart author of *Cinderella Ate My Daughter* and *Girls & Sex*, Orenstein . . . makes a convincing case for why parents who were aware of her work with girls and used to say they were relieved to have sons now 'realized that their job may actually be harder: they had to raise good men,' as she writes." —*Men's Health*

"[Joins] Orenstein's provocative 2016 hit *Girls & Sex* as another game-changer. . . . Candor, empathy, and easy humor . . . animate her reporting." —*San Francisco Chronicle*

"As a psychotherapist who's raising a boy, I can't think of a more important book for our times. Eye-opening and nuanced, this compassionate exploration of boys' sexual lives gives voice to their deepest struggles and should be mandatory reading for anyone who cares about the next generation—which is to say, all of us."

—Lori Gottlieb, *New York Times* bestselling author of *Maybe You Should Talk to Someone*

"[Orenstein] trains her expert eye on the world of adolescent boys and the unique set of challenges that young men are facing today. *Boys & Sex* is not just a candid and often devastating view into the lives of real high school and college boys right now; it's an affirmation of hope and an exercise in the power of listening." —*Salon*

"Orenstein's journey into the world of the young American male may not reassure . . . but it's essential." —*People*

"A sobering look at the landscape in which young men are growing up—and an invitation for the grown-ups in their lives to offer a lot more support and direction." —*Chicago Tribune*

"Peggy Orenstein has done something rare. She has listened to young men in ways that have allowed them to speak candidly about the fraught world of their sexuality, and she has been true to the complexities of their experiences—their hopes but also the fears, shame, pressures, and angers that cause them to violate others and corrode their capacity for care and love. What they say is scary and heartbreaking and vitally important for us all to hear. This is a bracing, insightful, humane, engaging, invaluable book. And it charts the course for real change."
—Richard Weissbourd, Senior Lecturer and Faculty Director of Making Caring Common, Harvard Graduate School of Education

"Expertly written. . . . [A] candid and fascinating portrait of young American masculinity." —*Publishers Weekly*, starred review

"Masculinity doesn't have to be toxic—but so much of what we communicate to boys about masculinity is just that. The amazing Peggy Orenstein does more than just document the atrocities. She listens to boys and bears witness to their efforts to free themselves from the trap the culture sets for them. *Boys & Sex* is required reading for anyone who has ever loved, raised, been, or will become a boy."
—Dan Savage, bestselling author and host of *Savage Podcast*

"Peggy Orenstein dared to do what so many of us are afraid of: actually ask boys about sex and then listen to what they had to say. She has given boys the opportunity to speak honestly about their feelings around sexuality, pornography, gender, consent, and so much more. Their answers are illuminating, oftentimes surprising—and essential." —Nick Kroll, cocreator, writer, and star of *Big Mouth*

"Forget what you thought you knew about boys and sex. Here, at last, is an honest book about the sexual lives of boys and young men; the good, the bad, the endlessly complicated and emotionally fraught. Peggy Orenstein has peeled back typical male bravado and exposed the raw hearts of boys struggling to navigate a confusing sexual landscape. *Boys & Sex* is a crucial contribution to the long overdue conversation about masculinity." —Michael Ian Black, author, comedian, and actor

"Essential reading." —PureWow

BOYS
& SEX

ALSO BY PEGGY ORENSTEIN

Don't Call Me Princess: Essays on Girls, Women, Sex, and Life

Girls & Sex: Navigating the Complicated New Landscape

Cinderella Ate My Daughter: Dispatches from the Front Lines of the New Girlie-Girl Culture

Waiting for Daisy: A Tale of Two Continents, Three Religions, Five Infertility Doctors, an Oscar, an Atomic Bomb, a Romantic Night, and One Woman's Quest to Become a Mother

Flux: Women on Sex, Work, Love, Kids, and Life in a Half-Changed World

Schoolgirls: Young Women, Self-Esteem, and the Confidence Gap

PEGGY ORENSTEIN

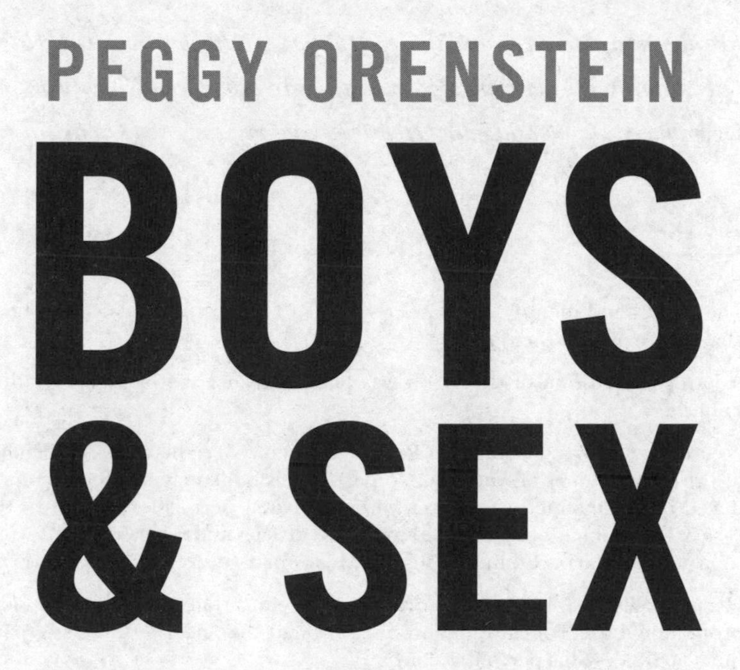

BOYS & SEX

YOUNG MEN ON HOOKUPS, LOVE, PORN, CONSENT, AND NAVIGATING THE NEW MASCULINITY

HARPER

NEW YORK • LONDON • TORONTO • SYDNEY

TO THE BEST OF MEN:

My husband, Steven Okazaki; my nephews, Matthew Orenstein, Harry Orenstein, and Mike Yamada; my brother-in-law, Geoff Kawafuchi; my big brothers, David and John Orenstein; and my dad, Mel Orenstein

HARPER

The names and identifying characteristics of certain individuals have been changed to protect their privacy.

A hardcover edition of this book was published in 2020 by HarperCollins Publishers.

FIRST HARPER PAPERBACKS EDITION PUBLISHED 2021.

The Library of Congress has catalogued the hardcover edition as follows:

Names: Orenstein, Peggy, author.
Title: Boys & sex : young men on hookups, love, porn, consent, and navigating the new masculinity / Peggy Orenstein.
Other titles: Boys and sex
Description: First edition. | New York : Harper, an Imprint of HarperCollins Publishers, [2020] | Includes bibliographical references and index.
Identifiers: LCCN 2019040572 (print) | LCCN 2019040573 (ebook) | ISBN 9780062666970 | ISBN 9780062666994 (ebook)
Subjects: LCSH: Teenage boys--Sexual behavior--United States. | Teenage boys--United States--Attitudes. | Young men--Sexual behavior--United States.
Classification: LCC HQ27.3 .O74 2020 (print) | LCC HQ27.3 (ebook) | DDC 305.235/10973--dc23
LC record available at https://lccn.loc.gov/2019040572
LC ebook record available at https://lccn.loc.gov/2019040573

ISBN 978-0-06-266698-7 (pbk.)

23 24 25 26 27 LBC 8 7 6 5 4

Contents

INTRODUCTION
What About the Boys? 1

CHAPTER 1
Welcome to Dick School 7

CHAPTER 2
If It Exists, There Is Porn of It 39

CHAPTER 3
Are You Experienced? Life and Love in a Hookup Culture 73

CHAPTER 4
Get Used to It: Gay, Trans, and Queer Guys 105

CHAPTER 5
Heads You Lose, Tails I Win: Boys of Color in a White World 135

CHAPTER 6
I Know I'm a Good Guy, but . . . 165

CHAPTER 7
All Guys Want It. Don't They? 183

CHAPTER 8
A Better Man 197

CHAPTER 9
Deep Breath: Talking to Boys 219

Acknowledgments 239
Notes 241
Bibliography 263
Index 281

BOYS
& SEX

INTRODUCTION

What About the Boys?

I never imagined I'd write about boys. As a journalist, I have spent a quarter of a century chronicling *girls'* lives. That has been my passion, my calling. But traveling the country in the wake of publishing my book *Girls & Sex* (which exposed the contradictions young women face in their intimate encounters), I was urged at every stop—by parents, by girls, by guys themselves—to turn my attention to young men. Still, I resisted. Girls' lives, after all, were the ones that had been transformed by feminism; their parents were the ones driven to demand further change. Our expectations of boys had modified some, but not nearly as much.

Then came #MeToo. And Harvey Weinstein. And Bill Cosby. And Louis C.K. And Kevin Spacey. And Matt Lauer. And Shitty Media Men. And Travis Kalanick. And Roy Moore. And a pussy-grabbing president. And the nude photo–sharing scandal in the US Marines. And Brock Turner. And Owen Labrie. And, for that matter, Aziz Ansari. And the viral *New Yorker* story "Cat Person." The pervasiveness, the sheer magnitude of sexual misconduct across every sector of society by men young and old became glaringly, disturbingly obvious. Masculinity was declared to be "broken" and "toxic." The parents of boys I met, the ones who when I described my work with girls used to shake their heads ruefully and say they were "relieved to have sons," suddenly realized that their job may actually be harder: they had to raise good men. Perhaps, I thought, this moment would be a breakthrough, offering not only a mandate to reduce sexual violence, but an opportunity to engage young men in authentic, long-overdue conversations about gender and intimacy.

Now I was interested, so I did a little digging. I already knew that Americans talk precious little to their daughters about sex, but I'd soon learn they talk even less to their sons. True, boys may now more likely be warned to "respect women," but what, precisely, does that mean? Which women, under what circumstances, and how? What's more, despite the growing insistence that only "yes" means "yes," boys (like girls) are bombarded by incessant images—on TV, in movies, games, social media, music videos—of female objectification and sexual availability, which are reinforced by unprecedented exposure to pornography. How were they navigating that dissonance? I had read lots of hand-wringing think pieces *about* boys but heard very little *from* them—their own voices were absent from discussions of their behavior in a way that girls', whose activism has pushed social change, were not. I'd told a crucial story in talking to young women, but I realized it wasn't enough. If I truly wanted to help promote safer, more enjoyable, more egalitarian, more *humane* sexual relationships among young people, I needed to go back into their world and have the other half of the conversation.

My biggest fear when I started my reporting was that guys wouldn't talk to me. Unlike girls, they don't exactly have a reputation for chattiness. Plus, I look like I could be their mom. But if anything, they were *more* forthcoming, talking extensively, honestly, bluntly, eagerly, including—perhaps especially—about the very thing boys are supposedly loath to discuss: their feelings. They confided their insecurities, pressures, and pain; their anxieties about sexual performance; their desire to connect and their fears about doing so. They talked about sexual pleasure—their own as well as their partners', and when they cared (or didn't) about the latter. They wrestled over the influence of porn; their feelings about casual hookups; how their race, sexuality, and gender identity affected their perceptions of masculinity. They struggled with the social cost of challenging "locker room talk." They chafed against the assumption that "guys only want sex."

Many of the boys I met treated our sessions as a protected space in which to reflect, to unburden themselves, sometimes to ask if they were "normal." Often, they paused during a conversation, drew a deep breath, and said, "I've never told anybody, but . . ." Or, "Fuck it, I'm just going to tell you this. . . ." That usually preceded an anecdote about a time they toppled from the "good guy" pedestal, a fall that didn't fit—and threatened to undermine—their self-image. Other times they disclosed worrisome high-risk behavior or their own sexual abuse. I was surprised by how raw our conversations could be, but boys so rarely feel permission to speak candidly, from the heart, about their interior lives.

I spent over two years talking to young men between the ages of sixteen and twenty-two, engaging in in-depth, hours-long conversations about masculinity as well as their attitudes, expectations, and early experiences with sex and intimacy. I recruited my subjects through high school teachers, counselors, and college professors I'd met while working on *Girls & Sex*; on campuses where I was asked to speak; and through girls whom I'd interviewed over the years. After we met, boys themselves sometimes introduced me to friends or roommates who had differing perspectives. To protect their privacy, I have changed their names and other identifying details.

I do not in these pages claim to reflect the experience of all young men. That would be unrealistic. The girls I wrote about in *Girls & Sex* were either in college or college bound: to ensure equivalency, so were the boys. The young men of color, in particular, were a distinct group: attending predominantly white schools, conducting their social lives in that world, subject to its specific forms of gendered racism. Also, since my interest lies in the mainstream, I did not wade deep into the muck of the manosphere: I believe, however, that we can learn something about the rage of the incels, MGTOW, Jordan Peterson fans, mass shooters, and other extremists by listening to the ideas, assumptions, and pressures of ordinary boys. Beyond that, though, I cast my net wide, talking to guys from every region of the country, from big cities and small towns, attending both public and private high schools and colleges. They spanned a spectrum of race, religion, and, to a degree, class. They were gay, straight, bisexual, and transgender (an identity I did not explore in *Girls & Sex* but was more conscious of this round). Although they leaned toward the politically progressive, they were not exclusively so. Many were athletes. Some were in fraternities; some had disaffiliated over the treatment of women by their "brothers." Some watched porn daily; a few renounced it entirely. Plenty admitted to having treated girls shabbily. Some acknowledged engaging in sexual misconduct—even if they had not, though, they all knew a guy (usually more than one) in high school or college who had: sometimes that person had been a close friend. They grappled with how to demand accountability from those boys, how to take it on themselves.

All in all, I spoke with far more boys than I had girls: well over one hundred. That was, in part, intentional—I wanted to be thorough, given that I hadn't previously written about young men—but it was also significantly easier to obtain permission to talk to them, notably from parents of those who were underage. Maybe adults were simply less protective of their sons, but I suspect some were also hoping that I would not only interview their boys about sexual behavior, but educate them as well, saving parents the awkwardness of doing so themselves.

Would I have gotten a different story from these young men had I been male? I can't say. I came to believe, though, that being a woman had its advantages. I don't know that boys would have been as emotionally open with a man. At the very least, I'd say that for anything they withheld because I was female, there was something else they expressed for that precise reason. They often commented on the disconnect between our discussions and the ways they'd learned to talk with other guys about sexual encounters—mainly as a means to shore up masculinity. Away from that pressure, they could acknowledge its negative impact on their mental health without seeming weak or feeling judged. Perhaps that's why some young men would *insist* I interview them, emailing me out of the blue ("I heard you say on the radio you were writing a book about boys. . . ."), corralling me after a speaking engagement, or checking in multiple times to set a date if I didn't respond quickly. More stayed in touch than I would have imagined, too, texting or emailing for advice on complicated situations in which they found themselves. For a few, those relationships continue; at this point, as unlikely as it seems, we have become friends.

Had I encountered these guys in their daily lives—had I been their mother, their aunt, their teacher—I would never have been privy to their innermost thoughts. They trusted me with that access because, as much as I did, they wanted me to get it right. The good news is, there was a deep desire among many of them for something different: a more expansive, holistic version of masculinity; a hunger for more guidance about growing up, hooking up, and finding love in a new era. In order to provide that, though, the first step is to listen to what they have to say.

CHAPTER 1

Welcome to Dick School

I knew nothing about Cole before meeting him; his was just a name on a list of boys who'd volunteered to talk to me (or perhaps had their arms twisted a bit) through a counselor at an independent high school outside of Boston. The afternoon of our first interview I was running late. As I rushed down a hallway I noticed a boy sitting outside the library waiting—it had to be him—staring impassively ahead, both feet planted on the floor, hands resting loosely on his thighs.

My first reaction was, *Oh no.*

It was totally unfair, a breach of journalistic objectivity, a scarlet letter of personal bias. Cole, eighteen, would later describe himself to me as a "typical tall, white athlete guy," and that is exactly what I saw: he topped six feet, with broad shoulders and short-clipped, dirty blond hair. His neck was so thick that it seemed to merge right into his jawline. His friends, he would tell me, were "the jock group. They're what you'd expect, I guess. Let's leave it at that." What's more, he was planning to enter a military academy for college the following fall. If I had closed my eyes and described the boy I imagined would *never* open up to me, it would have been him.

But Cole surprised me. He pulled up a picture on his phone of his girlfriend, whom he'd dated for the past eighteen months, describing her proudly as "way smarter than I am," a feminist, and a bedrock of emotional support. He also confided how he'd worried four years ago, during his first weeks as a freshman on scholarship in a new community, that he wouldn't know how to act with other guys, wouldn't be able to make friends. "I could talk to *girls* platonically," he said. "That was easy. But being around guys was different. Because I needed to be a 'bro,' and I didn't know how to do that."

Whenever Cole uttered the word "bro," he shifted his weight to take up more space, rocked back in his chair, spoke low in his throat like he'd inhaled a lungful of weed. He grinned when I pointed that out. "Yeah," he said, "that's part of it: seeming relaxed and never intrusive, yet somehow bringing out that aggression on the sports field. Because a 'bro'"—he rocked back again—"is always, always an athlete."

Cole eventually found his people on the crew team, though it hadn't always been an easy fit. He recalled an incident two years prior when a senior was bragging in the locker room that he'd convinced one of Cole's female classmates—a *young* sophomore, Cole emphasized—that they were an item, then started hooking up with other girls behind her back. And he wasn't shy about sharing the details. Cole and another sophomore told the guy to knock it off. "I started to explain why it wasn't appropriate," Cole said, "but he just laughed."

The following day, a second senior started talking about

"getting back at" a "bitch" who'd dumped him. Cole's friend spoke up again, but this time Cole stayed silent. "And as this continued to happen," Cole said, "as I continued to step back and the other sophomore continued to step up, you could tell that the guys on the team stopped liking him as much. And they stopped listening to him, too. It's almost as if he spent all his social currency trying to get them to stop making sexist comments. And meanwhile, I was sitting there"—Cole thumped his chest—"too afraid to spend any of mine, and I just had buckets left.

"I don't know what to do," he continued earnestly. "Once I'm in the military, and I'm a part of that culture, I don't want to have to choose between my own dignity and my relationship with others that I'm serving with. But"—he looked me straight in the eye—"how do I make it so I don't have to choose?"

You're a Bitch If You Talk About Feelings

Lightning Round: Boys, describe the ideal guy!

"Reserved. You can't flaunt your emotions. You have to be strong. Emotionally and physically. If I have issues, if I have something wrong, that's my problem. I have to deal with it."

—Tristan, eighteen, Los Angeles

"You've got to look ripped, be tall, have fair skin, talk to a lot of girls. Your basic stuff, pretty much. I don't fit into it at all, because for one, I'm Latino. And I'm short. And I'm not ripped."

—Marcos, sixteen, Hoboken

"Definitely a business major. Involved in Greek life. Graduating and getting a job at Morgan Stanley. Making, like, $250,000 a year. Working eighty hours. I mean, that's probably the goal."

—Chris, twenty, Raleigh

"You have to be smart but also 'hood,' or whatever you want to call it. Combine them just right and you're the perfect black guy."

—Taye, seventeen, Washington, DC

"The biggest single determining factor is assertiveness. If I am dominating other people, I am being masculine."

—Ryan, eighteen, San Francisco

"Competitive. Definitely competitive."

—Jason, twenty-one, Seattle

"Chill. You gotta be chill. Not take things too seriously."

—Zach, twenty, Portland

"You can handle yourself, you don't take disrespect."

—Jaylen, eighteen, Baltimore

"If you want to get girls, you've got to be mean. You've gotta be an asshole."

—James, sixteen, San Jose

"Stamina. You want to be able to say, 'Dude, I fucked her for hours.*'"*

—Michael, eighteen, San Francisco

"Sports is a big deal. If you're good at sports, you're okay as a guy. And hooking up with a lot of girls, definitely. Commitment is a sign of weakness."

—Oscar, seventeen, Boston

"Athletic. You're at every party, but not partying too *much. You're hooking up with multiple girls, but not* every *girl. You're smooth. You're social. You've got game."*

—Connor, twenty-one, Philadelphia

"Athletic."

"Athletic."

"Athletic. Definitely, athletic."

For over two years, I talked to boys—dozens of boys—from cities and towns across America. Nearly all of them held relatively egalitarian views about girls, at least in the public sphere: they considered their female classmates to be smart and competent; entitled to their place on the sports field and in school leadership; deserving of their admissions to college and of professional opportunities. They all had platonic female friends. That was a huge shift from what you might have seen fifty, forty, maybe even twenty years ago. Yet, when I asked them to describe the ideal guy, those same boys, who were coming of age in the 2000s, appeared to be channeling 1955; their definition of masculinity had barely budged. Emotional detachment. Rugged good looks (with an emphasis on height). Sexual prowess. Athleticism. Wealth (at least someday . . .). Dominance. Aggression. Like the girls I had interviewed a few years before, they were in a constant state of negotiation, trying to live out more modern ideas about gender yet unwilling or unable to let go of the old ones. And, also as with girls, the most insidious aspects of the "ideal" were reinforced or glamorized for them at every turn: on athletic teams, in media, even in the home. Nearly 60 percent of American boys in an international 2017 survey said that their parents (usually their dads) were the primary source of restrictive messages about masculinity. Rob, an eighteen-year-old from New Jersey in his freshman year at a North Carolina college, said his father would tell him to "man up" if he was struggling in school or on the baseball field. "That's why I never talk to anybody about any problems I'm having," he said. "Because I always think, *If you can't handle this on your own, then you aren't a man, you aren't trying hard enough, you're being a bitch.*" Rob's roommate, Ely, who grew up in the suburbs of Washington, DC, got a similar message, though in a subtler form. "My dad wasn't sexist," he said. "I didn't learn 'toxic' or homophobic behavior from him. But I certainly learned the emotionally stunted side of masculinity. He never showed emotion: he was more of a sigh-and-walk-away guy than someone who would talk to you about what was going on."

A 2018 national survey of over a thousand adolescents found that although girls believed there were "many ways to be a girl" (the big, honkin' caveat being they still felt valued primarily for their appearance), boys felt there was only one narrow pathway to successful manhood. They still equated the display of most emotions, as well as vulnerability, crying, or appearing sensitive or moody, with "acting like a girl"—which, in case you were wondering, is not a good thing. A third of the boys surveyed agreed that they should hide or suppress their feelings when they were sad or scared. Another third, like Rob, had felt pressure to "be a man" and "suck it up." Over 40 percent agreed that when they were angry, society expected them to be aggressive; the next most common response was that they should do nothing, keep quiet, and, again, "suck it up." Only 2 percent associated maleness with qualities such as honesty or morality, and only 8 percent with leadership—traits that are, of course, admirable in anyone but have traditionally been ascribed to masculinity. Young American men also report more social pressure than other nationalities to be ever ready for sex and have as many partners as possible; feel a stronger stigma against homosexuality; and receive more messages that they should conform to rigid gender roles in the home and control their female partners, even to the point of violence.

Feminism may have afforded girls an escape from the constraints of conventional femininity, offered them alternative identities as women and a language with which to express the myriad problems-that-have-no-name, but it has made few inroads with boys. Whether you label it the "mask of masculinity," "toxic masculinity," or "the man box," the traditional conception of manhood still holds sway, dictating how boys think, feel, and behave. Young men who most internalize masculine norms (though which, at least to some extent, do not?) are six times more likely than others both to report having sexually harassed girls and to have bullied other guys. They are also more likely to have *themselves* been victims of verbal or physical violence (including murder). They are more prone to binge-drinking and risky sexual behavior, and more likely than other boys to be in car accidents. They are also painfully lonely: less happy than other guys, with fewer close friends; more prone to depression and suicide. Whatever comfort, status, or privilege is conferred by the "real man" mantle, then—and clearly those exist—comes at tremendous potential cost to boys' physical and mental health, as well as that of the young women around them.

At its core, what psychologist William Pollack calls "the boy code" trains guys to see masculinity in opposition to, and adversarial toward, femininity: a tenuous, ever-shifting position that must be continuously policed. Anything that smacks of "girlieness"—in oneself, in other boys, and, of course, as embodied by actual girls—must be concealed, ridiculed, or rejected. Love, connection, and vulnerability are signs of weakness; aggression is celebrated and eroticized; conquest is everything. That fear of appearing subordinate to other guys, according to Cole, the boy I met outside the high school library near Boston, was one of the reasons he preferred to partner with girls on school projects. "There's more risk involved with working with another guy," he explained. "With a guy, I need to act as unapproachable as possible. It's like, if I come off as too communicative, I'll be the underling. When I'm working with a girl, it feels safer to talk and ask questions, to work together, or to admit that I did something wrong and that I want help."

For Cole, as for many boys, the rules and constraints of masculinity were a constant presence, a yardstick against which all their choices were measured. Once, during his junior year, he suggested that his teammates go vegan for a while, just to show that athletes could. "And everybody was like, 'Cole, that is the dumbest idea ever. We'd be the slowest in any race.' And that's somewhat true: we do need protein. We do need fats and salts and carbs that we get from meat. But another reason why they all thought it was stupid is because being vegans would make us *pussies*."

Cole was raised mostly around his mother, grandmother, and two younger sisters—his parents divorced when he was ten, and his dad, who was in the army and then the National Guard, was often away on active duty. Cole spoke of his mom with unequivocal love and respect. His father was another matter. On one hand, the older man was a caring and present dad, even after the split with Cole's mom. But the stoic expression that had made me leery? Cole got that from him. "I find it hard to be emotionally expressive," he said. "Especially around my father. He's a nice guy. But I can't be myself around him. I feel like I need to keep everything that's in here"—Cole thumped his chest again—"behind a wall, where he can't see it. It's a taboo, like . . . not as bad as incest, but I feel like that's a part of me that I'm not supposed to show."

It may seem extreme to equate a discomfort with emotions to incest, but I understood what Cole meant. He wasn't the first boy to use the phrase "behind a wall" when referring to feelings. Noah, a sophomore at a Los Angeles–area college, said he, too, built "a wall" to conceal any emotional vulnerability: "I felt it happen starting in high school," he recalled. "And it's not like my dad is some alcoholic, emotionally unavailable asshole-with-a-pulse. He's a normal, loving, charismatic guy who's not at all intimidating. And my group of friends—I don't mean to put it on these guys because they're still my best friends, the best people I've met in my life. Maybe they would say they had to act that way as much because of me as I say I did because of them. But there's a block there. There's a hesitation that I don't like to admit. A hesitation to talk about . . . anything, really. We learn to confide in *nobody*. So you sort of train yourself not to feel."

There is no difference at birth between boys' and girls' need for connection, nor, neurologically, in their capacity for empathy—there's actually some evidence that infant boys are the more expressive sex. Yet, from the get-go, they are relegated to a more restrictive emotional landscape. In a classic study, adults shown a video of an infant startled by a jack-in-the-box were more likely to presume the baby's response was "angry" if first told the child was a male. Mothers of young children have repeatedly been found to talk more to their daughters and to employ a broader, richer emotional vocabulary. With sons, again, they focus primarily on one emotion: anger. (Fathers speak with less emotional range than mothers regardless of their child's gender, though they do sing to and smile at their daughters more often, as well as more readily acknowledge girls' sadness.) Despite that, according to Judy Y. Chu, who studies early gender socialization, preschool boys retain a keen understanding of feelings and a desire for close relationships. But by midway through kindergarten—that's age five or six—they've learned from their peers to knock that stuff off, at least in public: to disconnect from feelings, shun intimacy, and become more hierarchical in their behavior. The lifelong physical and mental health consequences of that gender performance are ingrained as early as age ten. By fourteen, boys become convinced that other guys will "lose respect" for them if they talk about problems (early adolescence, incidentally, is both critically important to boys' development and vastly understudied). They suspect girls won't be attracted to them, either, and, frankly, they may be right: in a Canadian study, college-aged women found men who used shorter words and spoke less to be more appealing than others. And Brené Brown, who calls vulnerability the special sauce that holds relationships together, has noted that even women who *claim* to want guys who are emotionally transparent may grow uncomfortable around, or even reject, men who respond.

"Emodiversity"—being able to experience a broad sweep of emotions, positive as well as negative—is crucial to adults' emotional and physical health. Yet, in our conversations, boys routinely confided that they felt denied—by parents, male peers, girlfriends, media, teachers, coaches—the full gamut of human expression, especially anything related to sorrow or fear. For Rob, the college freshman in North Carolina, as for many of the guys I spoke with, our interview was a unique opportunity to grapple openly with his experience not only of sex but of emotional intimacy. We met in January, about four months after he'd broken up with his high school girlfriend. The two had dated for over three years—"I really did love her," he said—and although the colleges they now attended were thousands of miles apart, they'd decided to try to stay together. Then, in late September, Rob heard from a friend that she was cheating on him: she'd hooked up with multiple guys at parties and had intercourse with at least one. "So I cut her off," he said, snapping his fingers. "I stopped talking to her and forgot about her completely." Only . . . not really. Although he didn't use the word, the reality was that Rob spiraled into depression: the excitement he'd felt about leaving home, starting college, rushing a frat all drained away, and, as the semester wore on, it didn't come back.

I asked who he talked to during that time.

"That's the problem," Rob said. "None of my friends talk about feelings. If you were hung up over a girl, they'd be like, 'Stop being a bitch.'" Rob looked glum. He'd never even confided in his best friend, whom he met playing baseball in eighth grade. The only person with whom Rob could drop his guard, be fully open, was his girlfriend, but that was no longer an option—not when she was the source of his pain. So, although he tried to ignore his grief, it only festered.

Girlfriends, mothers, and, in some cases, sisters were the most commonly cited confidants among boys I met, and while it's wonderful to know they have *someone* to talk to—and I'm sure mothers, in particular, savor the role—teaching boys that women are responsible for emotional labor, for processing men's emotional lives in ways that would be emasculating for guys to do themselves, comes at a price to both sexes. Among other things, that dependence can leave boys stunted, in a state of arrested development, potentially unprepared to form caring, lasting, intimate relationships.

Rob did, eventually, schedule a session with his campus counseling service, but he only went once. "I just sat there the whole time and didn't say anything," he said. "It felt too weird to openly discuss feelings."

I asked him to expand on the word "weird." "It went back to making me seem like I was weak. Like, 'I can get through this. I'm fine.'"

"But were you fine?" I asked.

"No," he admitted. "I wasn't." By Thanksgiving break, he was so despondent that he had what he called a "mental breakdown" after dinner one night while chatting in the kitchen with his mom. "I was so stressed out," he said. "Classes. The thing with my girlfriend. It was a lot for me to handle." He couldn't describe what that "breakdown" looked or felt like to him (though, he said, it "scared the crap" out of his mom, who immediately demanded, "Tell me everything"). All he could say definitively was that he didn't cry. "Never," he insisted. "I don't cry *ever.*"

I paid close attention when boys mentioned crying—doing it, not doing it, wanting to do it, not being able to do it. For most, it was a rare and sometimes shameful event—a dangerous crack in that carefully constructed internal edifice. True, teary men are given more leeway today than they once were, but there are rules. *GQ*, for instance, stipulates that guys can cry in extreme pain: "Like, say, if a piano were dropped from a fifth-story window onto your foot"; if someone is shooting at you; if you're an athlete and your team wins a championship (LeBron after the 2016 NBA finals); or at a tear-jerking film (that "is a bodily function . . . like menstruation of the eyes"). Askmen.com adds that crying over a sports loss is acceptable—providing you weren't responsible for it (so, the site says, Tim Tebow after losing the 2009 SEC Championships is fine, but *not* Roger Federer after blowing the Australian Open that same year). A survey of 150 college football players found that while they believed it was reasonable to "tear up" after an important victory or a devastating loss, actual sobbing, for most, was never an option.

A college sophomore in Chicago told me that he hadn't been able to cry when his parents divorced. "I really *wanted* to," he said. "I *needed* to cry." His solution: he streamed three movies about the Holocaust back to back (that worked). As someone who has, by virtue of my genitalia, always had permission to weep, I didn't initially get it. It took multiple interviews for me to realize that when boys confided in me about crying—or, even more so, when they teared up right in front of me—they were taking a risk, trusting me with something private and precious: evidence of vulnerability, or a desire for it. Or, as with Rob, an inability to acknowledge it that was so poignant, it made me want to, well, cry.

Jocks, Bros, Fags, and Pussies

It was easy for boys to reel off the excesses of masculinity—they, too, had seen headlines about mass shootings, domestic violence, sexual harassment, campus rape, presidential Twitter tantrums, and Supreme Court confirmation hearings. Even a football player at a Big Ten school bandied about the term "toxic masculinity" in our conversation ("Everyone knows what that is," he said, when I seemed surprised). They had more difficulty with another question: what they *liked* about being a boy. "Huh," said Josh, a college sophomore in Washington state. "That's interesting. I never really thought about that." After some hemming and hawing, he landed on sports—"I *love* playing sports," he said, but then again, so did his younger sister, so maybe that wasn't a gender thing. Still, athletics—the physicality, the camaraderie, the competition, the healthy release of aggression, the pure delight in the game—was the most frequent response boys came up with, and for many what defined boyhood for them. They recalled their early days on the playing field with nostalgic, almost romantic, warmth. Then something changed and the very thing that felt sustaining became, for many, oppressive. "I've played lacrosse for almost twelve years," said Eric, a high school senior in Los Angeles. "And I love it. But what I *don't* love is the egotistical, 'I'm good, and I'm an asshole' culture."

I was struck by how many boys told me they'd dropped out of sports they enjoyed not because they didn't have the skill to continue, but because they couldn't stand the *Lord of the Flies* mentality of teammates or coaches. "I played soccer up until my junior year of high school," said Noah, the Los Angeles–area college sophomore. "I always loved the sport, but being on a team became way too 'bro-y' for me. There was always that need to prove how tough, how manly, how careless you were. It made me feel like an outsider."

Perhaps most extreme was Ethan, originally from the Bay Area, who had been recruited by a small liberal arts college in New England to play lacrosse. "I knew there would be that East Coast, 'lax bro' culture," he said, "but I didn't know how intense it would be. It was all about sex and bragging about hooking up, and even the coaches endorsed victim-blaming. And everyone used 'fag' really freely. They weren't like that in their classes or around other people; it was a super-liberal school. But once you got them in the locker room . . ." He shook his head. As a freshman, Ethan didn't feel he could successfully challenge his older teammates, especially without support from the coaches. So he quit the team; not only that, he transferred. "The school I was at was so small," he explained. "If I'd stayed, there would've been a lot of pressure on me to play, a lot of resentment, and I would've run into those guys all the time. . . . This way I didn't really have to explain anything." At his new school, Ethan added, he no longer played any sports.

Robert Lipsyte, a longtime sports journalist, believes that it is boys' very joy in athletics that allows the conditioning of "jock culture" to take hold. Sports can teach the kind of courage, cooperation, grace, and grit that lay the groundwork for off-the-field success. Plus, you know, they're *fun*. But, Lipsyte has written, "jock culture" uses those values as a smoke screen for bullying, entitlement, aggression, violence, and a "win-at-all-costs attitude that can kill a soul."

Jock culture (or what the young men I met were more likely to call "bro culture") is the dark underbelly of male-dominated enclaves, whether or not they formally involve athletics: all-boys' schools, fraternity houses, Wall Street, Silicon Valley, Hollywood, the military. Even as they promote bonding, preaching honor and integrity, such groups condition guys to treat anyone who is not "on the team" (a category that may include any woman who is not a blood relative) as the enemy—bros before hos!—justifying hostility or antagonism toward them. Loyalty is unconditional, and masculinity asserted through sharing sexual exploits, misogynist language, and homophobia.

Cole had loved his time on crew and as a senior was made team captain. He relished being part of a unit, a band of brothers. When he raced, he imagined himself pulling each stroke for the guy in front of him, pulling for the guy behind him—never for himself alone. But not everyone could muster such higher purpose. "Crew demands you be pretty determined," Cole said. "You're pushing yourself to a threshold of pain, and you have to keep yourself there. And it's hard to find something to motivate you to do that other than anger and aggression. That's always reliable. I hear the music my teammates listen to before a race. It's all pretty violent stuff."

I asked him more about how those teammates talked in the locker room. That question always made guys squirm. They would rather talk about porn use, erectile dysfunction, premature ejaculation—*anything* but admit to a woman the truth about "locker room banter." Cole cut his eyes to the side, shifted in his seat, sighed deeply. "Okay," he finally said. "So, here's my best shot. We definitely say 'fuck' a lot; 'fuckin'' can go anywhere in a sentence. And, we call each other pussies. We call each other bitches. We never say the *N*-word, though. Because that's going too far."

"What about 'fag'?" I asked.

"No," he said, shaking his head firmly.

"So," I said, "why can't you say 'fag' or the *N*-word but you can say 'pussy' and 'bitch'? Aren't those just as offensive?"

"One of my friends said we probably shouldn't say those words anymore, either," he said, "but what would we replace them with? We couldn't think of anything that bites as much."

"Bites?"

"Yeah. It's like . . . for some reason, 'pussy' just works. When someone calls me a pussy—'Don't be a pussy! Come on! Fuckin' go! Pull! Pull! Pull!'—it just flows. If someone said, 'Come on, Cole, don't be weak! Be tough! Pull! Pull! Pull!' it wouldn't get inside my head the same way. I don't know why that is. I don't know why it needs to be those words." He paused a moment. "Well," he said, "maybe I do. Maybe I just try not to dig too deeply into that area because I know it's wrong."

Although losing ground in the most progressive circles, the use of "fag" remained pervasive among the boys I interviewed, especially during middle and high school. "That's so gay" had also long since displaced the out-of-favor "retarded" as a catchall for anything stupid, boring, irritating, disappointing, unpleasant, lame: your phone was "gay" if didn't have service, your pencil was "gay" if it broke, it was "so gay" for a movie to be sold out. "Fag," by contrast, was reserved specifically for other boys (never girls, nor do girls often malign one another with homophobic slurs, though they may use "dude" or "bro" to show affection). Rob, who had cut back on using "fag" once he got to college because "some people don't care, but other people take offense," said that he thought of it as "a universal term. Like, 'Oh, you're being annoying—stop being a fag!' Or, 'Oh, you're doing something weird—stop being a fag!' Or, let's say we were busting each other's chops and one of my friends takes it too seriously. Then it's like, 'Oh, stop being a fag! Stop being gay! Man up! We're joking!' "

Rob quickly added that none of his friends would've used the word in reference to *actual* homosexuals. "Never!" he said, aghast. "One of my closest friends came out our senior year of high school!"

Most boys today have no problem with gay *people*: they're supportive of LGBTQ+ rights and same sex marriage. At the same time, "fag" is the worst thing they can be called; it has become less a comment on their sexual orientation than a statement about their manhood. "Fag," according to C.J. Pascoe, a professor of sociology at the University of Oregon, is what draws the lines of the "man box," provides its essential contours. Much like "slut" for girls, its definition is fluid, elusive, which only intensifies its power: "fag" keeps guys perpetually vigilant (though it's not always clear against what) and shuts down any challenge or objection to the "boy code." Guys can be called "fag," Pascoe said, "for literally anything. Like dropping a piece of meat out of a sandwich." One boy I met was labeled "fag" in middle school because he liked to read (so he stopped). Another worried he'd be deemed a "fag" for not being up on his drug terminology, such as that a "moke" was a bong hit that mixed tobacco with marijuana. "I'm more embarrassed to get something wrong and seem dumb in front of a guy than I am in front of a girl," he told me. "I don't know why that is." Although he didn't connect the dots himself, "fag" also tacitly regulated Rob, kept him from talking to other guys about losing his girlfriend.

Despite its amorphousness, Pascoe found there were certain behaviors that most reliably provoked the epithet: showing emotion, being openly affectionate with other boys, behaving romantically toward a girl (which was seen as heterosexual in the "wrong" way and explained why one high school junior told me that having a girlfriend was "gay"), or appearing in any way incompetent.

Recently, Pascoe turned her attention to "no homo," a phrase that gained traction in the 1990s, sifting through over a thousand tweets, primarily by young men, that included the phrase as a hashtag. Most were about expressing a positive emotion, sometimes as innocuous as "I love chocolate ice cream, #nohomo" or "I loved the movie *The Day After Tomorrow,* #nohomo." "A lot of times they were saying things like 'I miss you' to a friend or 'We should hang out soon,'" she said. "Just normal human expressions of joy or connection. But they had to add #nohomo to inoculate themselves against other guys lobbing insults at them, to create a space where they could express those sentiments. So it became not just a homophobic joke, but also a shield that allowed them to be fully human."

Just because some young men now draw the line at referring to someone who is openly gay as a "fag" doesn't mean such boys (or those who have traits that read as "gay") are suddenly safe. If anything, gay boys were more conscious of the rules of manhood than their straight peers—they had to be. In fact, I began to joke that the gay boys I interviewed were like spies in the house of hypermasculinity.

Mateo, seventeen, attended the same Boston-area high school as Cole, also on a scholarship, but the two could not have presented more differently. Mateo was slim and tan (his father was Salvadoran), with an animated expression and a tendency to wave his arms extravagantly as he spoke. Where Cole sat straight and still, Mateo crossed his legs at the knee and swung his foot, propping his chin on one hand.

This was Mateo's second private high school; the oldest of six children, he had been identified as academically gifted and encouraged by an eighth-grade teacher to apply to an all-boys' prep school for his freshman year. When he arrived on campus, he discovered that his classmates were nearly all white, all athletic, all affluent, and, as far as he could tell, all straight. Mateo—Latinx and gay, the son of a janitor—was none of those things. He felt immediately conscious of how he held himself, how he sat, and especially of the pitch of his voice. He tried lowering it, but that felt unnatural, so he avoided conversation entirely. He changed the way he walked as well, so as not to be targeted as "girly." "One of my only friends there was gay, too," he said, "and he was out. He got *destroyed.*"

Across the country in San Francisco, Adam, twenty, a college sophomore who was also openly gay, insisted that he hadn't been offended by his high school classmates' use of "that's so gay" or "fag": "I wouldn't think, *That person hates gay people,*" he said. "My assumption was that he was joking around. That didn't necessarily make it *better*, but it didn't feel threatening." Maybe. Then again, despite growing up in what is arguably the gay capital of America, Adam, too, at that time, was in the closet. And like Mateo, he was forever monitoring himself—the way he sat, his hand gestures, his vocal inflection—in an attempt to evade detection. "I would practice my facial expressions in the mirror," he recalled, "especially my 'sit back and look like "the shit"' face. Then no one would question me. I was the man. I was *a* man." The only time Adam said he "slipped up" was during his first middle school mixer. "I was out there on the dance floor doing my thing, and this other boy came up to me and said I was using my hips too much, I was dancing like a girl. It was horrible to have something that I was naturally inclined to do be brought to my attention as 'wrong.' That was exactly the kind of thing that I was trying to avoid at all costs. So from then on, I would watch how other boys danced and do what I saw them doing."

Guys who identified as straight but weren't athletic, or were involved in the arts, or had a lot of female friends, all risked having their masculinity impugned. What had changed for this generation, though, was that some of them, particularly if they had grown up around LGBTQ+ people, didn't much care. "I don't mind when people mistake me for being gay," said Luke, a high school senior from New York City. "It's more of an annoyance than anything, because I want people to believe me when I say I'm straight." The way he described himself did, indeed, tick every stereotypical box—except the only one that counted: an interest in sex with other boys. "I know I'm not a big, masculine guy," he said. "I'm a very thin person. I like clothing. I care about my appearance in maybe a more delicate way. I'm very in touch with my sensitive side. I like to talk about emotions. So when people think I'm gay"—he shrugged—"it can feel like more of a compliment. Like, 'Oh, you like the way I dress? Thank you!'"

One of Luke's friends, who had been labeled "the faggot frosh" in ninth grade, was not so philosophical. "He treats everything as a test of his masculinity," Luke said. "Like, once I made a joke about his alcohol tolerance and he goes, 'Dude, I could drink you under the table!' And once when I was wearing red pants, I heard him say to other people, 'He looks like such a faggot.' I didn't care, and maybe in that situation no one was really harmed, but when you apply that attitude to whole populations you end up with Donald Trump as president."

IF EMOTIONAL SUPPRESSION and disparagement of the feminine are two legs of the stool that supports "toxic masculinity," the third is bragging about sexual conquest. In the now-notorious video of an interchange on an *Access Hollywood* bus, leaked a month before the 2016 presidential election, then-candidate Donald Trump bragged about forcibly kissing women ("I don't even wait"), claiming that when you're "a star, they let you do it. You can do anything. Grab 'em by the pussy." He later dismissed his comments as "locker room banter": the way men talk when they're alone. Professional athletes took to Twitter to express outrage at that characterization (Kendall Marshall, then in the NBA, tweeted, "PSA: sexual advances without consent is NOT locker room talk"; mixed martial artist CM Punk quipped, "'Grab them by the pussy' isn't talk from any locker room I've ever been in. It's a Ted Bundy quote."); Trevor Noah, host of *The Daily Show,* pointed out the difference between "saying dirty words and glorifying nonconsensual contact." The overall implication was that sure, yeah, guys talk about sex, but they don't high-five one another about *assault.* And perhaps they don't, not literally, but I'm not so sure that, in spirit, Trump's assertion was inaccurate.

Near the beginning of the school year, a classmate boasted about hooking up with one of Mateo's friends, referring to her as an "ugly bitch" and a "troll" (which would seem like reasons *not* to have sex with someone). "She was *begging* for it, dude," the boy crowed. "I totally *destroyed* her. I ripped her up."

Mateo was irate, but he didn't tell his female friend about it. What was the point? The comments weren't exceptional. "That's just the way straight guys talk. It's not like they say, 'Dude, I made her feel *great*!' That never happens. It's always, 'Bro! I *slammed* her!'"

At first I found it inexplicable that boys used such violent words in reference to sex. Why would you be proud of being a lousy lover? If they were truly talking about *sex* in those situations, they might bring up pleasure, connection, finesse: they wouldn't weaponize it. But the whole point of "locker room banter" is that it's *not* actually about sex, and that, I think, is why guys were more ashamed to discuss it as openly with me as topics that were equally explicit. Those (often clearly exaggerated) stories are in truth about power: about asserting masculinity through control of women's bodies. And that requires—*demands*—a denial of girls' humanity. In mixed-sex groups, teenagers may talk about "hooking up" (which is already impersonal—if you want to make them gag, use the phrase "making love"), but when guys are on their own, it can be hard to tell if they've engaged in an intimate act or have just returned from a visit to a construction site. They nail, they pound, they bang, they smash, they slam, they hammer. They "hit that," they "tap that ass," they "tear her up," they "destroy" her. The truth is less important than the posturing itself: using symbolic aggression toward women to bond and validate their heterosexuality. Dismissing that as "locker room banter" denies the ways that language can desensitize and abrade boys' ability to see girls as people deserving of respect and dignity. And, in fact, by the time they are in college, athletes are three times more likely than other students to be accused of sexual misconduct or intimate partner violence. That puts such bluster in a different light.

I don't know that the kind of elite young men who attended school with Mateo and Cole—those already firmly on the path to power—are *more* inclined to objectify women, but they are certainly not less so. Over the course of three months in 2016, for instance, sex scandals broke out among athletic communities at three top-tier colleges. October brought news of a "tradition" among the boys of the Harvard soccer team: a "scouting report" in which they rated new recruits to the freshman girls' team based on their perceived hotness, assigning each a sexual position in addition to the one she held on the field. "Yeah . . . She wants cock," a guy wrote about one of his female classmates. Another girl was named "the hottest *and* the most STD-ridden." A month later, a campus news site at Columbia University obtained racist, sexist, and homophobic screenshots from a wrestling team group chat. They had referred to the school's female students as "ugly socially awkward cunts" who feel "entitled" (apparently they didn't have a full grip on that concept). They claimed that they would "run the town of any state school" where "every girl begs for the cock so hard." (Guys I met at elite colleges, incidentally, regularly disparaged those at state schools—believing, against all evidence, that their exceptional SAT scores and socioeconomic status precluded harassment and assault rather than merely better insulating them from consequences.) In December, it emerged that members of Amherst's cross-country team had circulated an email that included photographs of eight girls—referred to as "friends of Amherst XC"—accompanied by their supposed sexual histories and penchants. One girl was termed "a walking STD" and of another, the author wrote, "Everyone needs their meatslab." The email turned out to be one of a series dating back at least two years that included such comments as "Do Asians really have horizontal vaginas?" and "You know the girls at your high school who aren't that attractive or personable, so no one talks to them? Picture a college with 900 of them and you have our lovely liberal arts institution." In the spring of 2019, two fraternities at Swarthmore—one of the nation's most politically progressive campuses—were forced to "voluntarily" disband after student-run publications released hundreds of pages of "minutes" from their house meetings that included, among other things, discussions of a "rape attic" and the acquiring of roofies; "finger-banging" a member's ten-year-old sister; racist comments about sexual acts with African American and Asian American women; vomiting on women during sex; and admiration for a brother who was known for "creampies coming at you whether you like it or not" (translation: ejaculating into a woman sans condom regardless of whether she consents).

Repugnant, yes. Unusual? Not so much. I live in the Bay Area, a bastion of liberalism, where high school boys in the affluent town of Piedmont were busted in 2012 for engaging in a "fantasy slut league" in which female students were "drafted" and boys earned points for "documented engagement in sexual activities" (the practice had been going on for at least five years when it was discovered, and there have since been quashed attempts to revive it). In 2016, students from several Silicon Valley high schools were caught sharing nudes of female classmates without their consent on a public Dropbox account. And in my own uber-leftie town of Berkeley, two separate groups of high school boys—one in 2014 and the other in 2017—posted pictures of their female friends (or girls who believed themselves to be their friends) on social media accounts with captions detailing the sexual acts the girls would perform. A third group, never identified, ran an Instagram account called "THOTs of Berkeley High" (THOT, a synonym for "slut," is an acronym for That Ho Over There) where pictures of girls of color, some surreptitiously snapped in school hallways, were posted; again, they were captioned with sexual acts the girls reputedly had performed or would perform.

When caught out, boys typically claim that they thought they were being "funny," just joking around. And in a way that makes sense—if, that is, you have been denied full emotional expression, been trained to suppress empathy, and consider cruelty, "ribbing," or making demeaning sexual comments about women to be forms of bonding. Such "humor" may even, when left unexamined, seem like an extension of the gross-out comedy of childhood: little boys are famous for their fart jokes, booger jokes, poop jokes, barf jokes. It's how they test boundaries, understand the human body, gain a little social cred with other guys. But, as with sports, their glee in that can both enable and camouflage sexism. The boy who, at age ten, asks his friends the difference between a dead baby and a bowling ball may or may not find it equally uproarious, at sixteen, to share what a woman and a bowling ball have in common (you can Google it). He may or may not post ever-escalating "jokes" about women, or African Americans, or homosexuals, or disabled people on a group Snapchat. He may or may not send "funny" texts to friends about "girls who need to be raped" or think it's hysterical to surprise a buddy with a meme in which a woman is being gagged by a penis, her mascara mixed with her tears. He may or may not, at eighteen, scrawl the names of his hookups on the wall of his all-male dorm, as part of a yearlong competition to see who can "pull" the most women. Perfectly nice, bright, polite boys I interviewed had done each of those things. How does that happen? A fifteen-year-old on the East Coast who had been among a group of boys at his school suspended for posting over one hundred racist and sexist "jokes" about classmates on a group Finsta (a secondary, or "fake," Instagram account that is often more authentic than a curated "Rinsta" or "real" account) reflected, "The Finsta connected us and became very competitive. You wanted to make your friends laugh, but when you're not face-to-face, you can't tell if you'll get a reaction or not so you go a little further. You go one step beyond. So it was the combination of competitiveness and that . . . *disconnect* that triggered it to get worse and worse as time went by."

In order for gross, crude, sexual, or even slapstick humor to be funny to its audience, researchers have found, it has to succeed in two contradictory things: violating morals (that is, it has to be disgusting) while seeming harmless, detached from any true reality; certainly you can't feel concern or identification with its subject. That a dead baby joke would be a whole lot less funny if you first described in detail how the baby suffered, the grief of the parents, the horror of the funeral. So, in order for boys to believe any of these antics were amusing, they had to systematically ignore the humanity of the girls involved—and that is not harmless at all.

At the furthest, most disturbing end of that continuum, "funny" and "hilarious" become a defense against charges of sexual harassment, misconduct, or assault. Consider the boy from Steubenville, Ohio, who was captured on video joking about the repeated violation of an unconscious girl at a party by a group of his friends. "She is so raped," he said, laughing. "They raped her quicker than Mike Tyson." When someone off camera suggested that rape wasn't funny, he retorted, "Rape isn't funny—it's *hilarious*!" One of the boys from Maryville, Missouri, who assaulted the unconscious fourteen-year-old Daisy Coleman, a subject of the Netflix documentary *Audrie and Daisy*, told police that in the moment he thought what they were doing was "funny." The high school lacrosse players from an all-male Catholic prep school in Louisville, Kentucky, who circulated pictures of their assault of sixteen-year-old Savannah Dietrich (a case that gained international attention when their lawyers threatened to sue her for tweeting their names after a slap-on-the-wrist sentence) also described their behavior as "funny." Again, recall that in order for a morally reprehensible act to be seen as a joke, it has to be considered harmless by the perpetrators; they have to resist identification with the subject, ignore pain. Anything less makes you a *pussy*.

"Hilarious" is another way, under pretext of horseplay or group bonding, that boys learn to disregard others' feelings as well as their own. "Hilarious" is a safe haven, a default position when something is inappropriate, confusing, upsetting, depressing, unnerving, or horrifying; when something is simultaneously sexually explicit and dehumanizing; when it defies their ethics; when it evokes any of the emotions meant to stay safely behind that wall. "Hilarious" offers distance, allowing them to subvert a more compassionate response that could be read as weak, overly sensitive, or otherwise unmasculine. "Hilarious" is particularly troubling as a defense among bystanders—if assault is "hilarious," they don't have to take it seriously, they don't have to respond: there is no problem.

"Hilarious" makes sexism and misogyny feel transgressive, rebellious rather than supportive of an age-old status quo. It also puts boys' hearts and heads into conflict, silencing conscience: they may know when something is wrong; they may even know that true manhood—or maybe just common decency—should compel them to speak up. At the same time, they fear that if they do, they'll be marginalized or, worse, themselves become the target of other boys' derision. Masculinity, then, becomes not only about what boys do say, but about what they don't—or won't, or *can't*—even when they wish they could. It blocks them from considering women's points of view, hardens them against compassion. Psychologist Michael Thompson has pointed out that silence in the face of cruelty or sexism is how boys *become* men. Charis Denison, a youth advocate and sex educator in the Bay Area, put it another way: "At one time or another, every young man will get a letter of admission to 'dick school.' The question is, will he drop out, graduate, or go for an advanced degree?"

The Sound of Silence

Cole and his girlfriend, like most high school couples, broke up at the end of senior year. They were headed for colleges in different parts of the country, but that wasn't the only reason. She was feeling "used" in their relationship, she said, as if Cole was only interested in spending time with her if they were having sex. "I wasn't . . ." he started to tell me, then broke off and began again. "No guy is *trying* to be a user. But I can't deny that was what was going down. I'd tell her I couldn't hang out because I had morning crew practice, or that I had to do homework during lunch because the night before I'd watched YouTube videos instead of working. I mean, say it however you want, but I was putting YouTube videos in front of her.

"When we broke up, she said, 'You're a really nice person, Cole, but you do a lot of things for yourself.' And what I think she meant was: I'm not someone who wants to hurt anybody, but I also don't care enough to go out of my way for people. She was definitely right. And she talked about my dad. . . . He always has to be the 'good guy.' And he *is* a good guy—when it doesn't cost him much, when it's easy. He's good with the big gesture. And I'm kind of like that, too. But also"—Cole paused and took a deep breath—"he wasn't a good guy to my mom. And I hope when I have a wife someday I can be more like her than like him."

Although she did almost all of the talking, that conversation with his girlfriend was the most emotionally direct the two had ever had. "The last time I ever said anything emotional to someone was . . . I don't know," he said. "Maybe it was with her. But actually it might have been the last time I talked to you."

Cole and I had been catching up on FaceTime. He was midway through his freshman year in college, and I was checking in to see how he'd resolved the conflict between his personal values and those of the culture in which he found himself. Most of his classmates were male, as he'd expected, and there was a lot of what passed for friendly ribbing: giving each other "love taps" on the back of the head; blocking one another's paths, then pretending to pick a fight; grabbing each other's asses; pretending to lean in for a kiss. Giving someone a hard time, Cole said, was always "easy humor," but it could slide into something more troubling pretty quickly. When one of his dorm mates joked to another, "I'm going to piss on you in your sleep," for example, the other boy shot back, "If you do, I'll fucking rape you." For better or worse, Cole said, that sort of comment no longer jarred him.

Although he had been adamantly against the epithet "fag" when we met, Cole found himself using that more, too, reasoning, as other boys did, that it was "the equivalent of 'You suck' or 'You're lame.'" Yet at least one of his friends had revealed himself to be legitimately homophobic, asserting that being gay was un-American ("I didn't know that about him until *after* we became friends," Cole hastened to add). And Cole had not met a single openly LGBTQ+ student at the school. He certainly wouldn't want to be out in this environment if he were gay. Nor, he said, would he want to be Asian—the two Asian American boys in his dorm were ostracized, treated like foreigners; both were miserable.

I pointed out, gently, that being able to silently disapprove of others' bigotry or homophobia was a luxury conferred by his own race and sexuality; he'd once told me he hoped to be "braver than that." Cole nodded. "I do feel kind of like a cop-out for letting all the little things slide," he said. "It's a cop-out to not fight the good fight. But, you know, there was that thing I tried sophomore year. . . . It just didn't work. I could try to be a social justice warrior here, but I don't think anyone would listen to me. *And* I'd have no friends."

During our first conversation, Cole had told me that he'd decided to join the military after learning in his high school history class about the Mai Lai massacre: the infamous 1968 slaughter by US troops of hundreds of unarmed Vietnamese civilians along with the mass rape of girls as young as ten. "I want to be able to be in the same position as someone like that commanding officer and *not* order people to do something like that," he'd said. I'd been impressed. But given that noble goal, that ideal of ethical resistance, was a single failure in the attempt to call out sexism a reason to stop trying? I understood that it was hard, that it was uncomfortable, that it was risky. I understood the personal cost might be greater than the impact. I also understood that, developmentally, adolescents want and need to feel a strong sense of belonging. No one wants to be excluded. And no one wants to court physical or emotional harm. But if Cole didn't practice standing up, if he didn't find a way to assert his values and find others who shared them, who was he?

"I knew you were going to ask me something like that," he said. "I don't know. In this hypermasculine culture where you call guys 'pussies' and 'bitches' and 'maggots'—"

"Did you say 'maggots' or 'faggots?'" I interrupted.

"*Maggots,*" he said. "Like worms. So you're equating maggots to women and to women's body parts in order to convince young men like me that we're strong. To go up against that, to convince people that we don't need to put others down to lift ourselves up . . . I don't know. I would need to be some sort of Superman."

Cole fell silent for a moment, his expression morose. "I think maybe the best I can do is to just be a decent guy," he continued. "The best I can do is lead by example." He paused again, furrowed his brow, then added, "I really hope that will make a difference."

CHAPTER 2

If It Exists, There Is Porn of It

Mason, nineteen, met me in the lobby of his dorm at the Big Ten university where he was a sophomore, dressed in a thrift-store cardigan, gray sweatpants, thick-soled sneakers, and two different-colored argyle socks. I couldn't tell what the length or color of his hair might be—it was completely covered, as were his eyebrows, by a neon-green stocking cap onto which he'd doodled stars and moons with a Sharpie, along with a favorite quote by Jean-Paul Sartre: "Like all dreamers, I mistook disenchantment for truth." When I commented on his outfit, Mason looked down quizzically, almost as if he were surprised to be wearing anything at all. "I guess I'm kinda 'dadcore,'" he said mildly. At five foot ten and 190 pounds, some might say Mason had a bit of a dad bod as well, though he was significantly trimmer than a few years before, when he played defensive lineman for his suburban Milwaukee high school football team. "I used to be pretty fat," he admitted. "The coaches would tell you to eat as much fast food as you can. They're like, 'After practice, have your parents take you to McDonald's.'" Mason was so soft-spoken, so laid-back that I had a hard time imagining him on the gridiron. The truth was, although he joined his first team at age eight, he'd never loved the game—for one thing, he hated being hit. He only continued because his two older brothers had played, so he thought he was "supposed to"; after eleventh grade, he finally bailed. "I understand it's a boy thing to get all rugged and dirty and bruised up," he said, "and I'm down for rugged and dirty, but I don't like experiencing physical pain."

That summer, in part because he thought his weight was hurting his chances with girls, he lost forty pounds by restricting his daily food intake to a granola bar, a few pretzels, and a couple handfuls of nuts. He also began following YouTube fitness sensations such as Aziz Shavershian, a skinny-kid video gamer–turned-bodybuilder who has only become more popular since his death from heart failure at age twenty-two, amid rumors of anabolic steroid abuse; Vegan Gains, who once posted footage of his eighty-two-year-old grandfather mid–heart attack to illustrate the perils of eating meat; and Frank Yang, who claimed to have put on over thirty pounds of chiseled muscle in just three weeks without the help of pharmaceuticals. Fitness gurus are to boys what beauty and "thinspo" vloggers can be to girls: a source of entertainment, motivation, and, potentially, self-harm. For a while, Mason was obsessed with how-tos on creating six-pack abs (which in actuality are only achieved through dangerous loss of body fat). He also went through a phase of watching freak-show videos about synthol, an oil that, when injected into the biceps or calves, creates the illusion of impossibly enormous muscles. Sometimes he followed recommendations that went beyond the gym: when a favorite vlogger enthused about the mind-expanding benefits of psilocybin, Mason tracked down mushrooms from a dark net site that he described as "like Amazon, but for drugs."

Turning to the internet for purported expert advice, guidance, and information is second nature to millennials, so I was surprised when, midway through our conversation, Mason pulled out a purple flip phone. He'd met a guy at the Lollapalooza music festival the previous summer who had made the switch; the boy claimed chucking the smartphone kept him more present in real life, less dependent on technology. "He made the point that for our age group, whenever there's a moment of silence people take out their phone to avoid awkwardness. And he wanted to embrace that awkwardness because life's an awkward endeavor." The idea intrigued Mason; plus, he'd been reading a lot about how smartphones and social media were linked to anxiety and depression among today's teens. He'd had a brush with those himself, during his first semester of college. His then-girlfriend was three hundred miles away at another school, but they were in near-perpetual contact via text and Snapchat. They also kept track of each other through a location-sharing app. So when his girlfriend didn't return his texts late at night, Mason could see precisely how often she was at the dorm of the guy with whom he suspected she was hooking up. "I was checking compulsively," he said. "It's so enticing. You know it's bad for you. You know it's going to be negative. But you can't stop doing it. So . . ." He waggled his flip phone.

"Also," he added, almost as an aside, "it keeps me from watching porn."

Porn World Versus Real World

It's no secret that today's children are guinea pigs in a massive porn experiment. Whereas (mostly) boys of previous generations might have passed around a filched, soiled copy of *Playboy* or possibly *Penthouse*, today anyone with a broadband connection can instantly access anything you can imagine—and a whole lot of stuff you don't want to imagine—right on their phones, more or less anonymously and regardless of how many obstacles parents try to put in place. "Porn" is, of course, a broad category, ranging from "lovemaking" purportedly posted by real-life couples, to fetish videos to eroticized violence and racism. It can be exploitative or ethically produced; it can be feminist. As the classic meme "Rule #34" states, "If it exists, there is porn of it. No exceptions." For the purposes of this discussion, though, "porn" refers to the sexually explicit video clips that heterosexual teenagers most readily and easily obtain (I'll talk about gay porn later). Also, I am restricting myself to minors and college students; adults' porn use is a separate issue.

Let's toss in a couple more caveats. Such as: curiosity about sex is normal. Sexual urges are normal. Masturbation is not only normal but healthy and important to sexual development (though sanctioned for and practiced by far more boys than girls). Seeking out erotica and information about sex is normal, too. The big change for today's young people, and the reason for new concern, is the aforementioned rise of high-speed internet. More specifically, it was the advent of Pornhub. Launched in 2007, when the boys I interviewed were on the cusp of puberty, Pornhub, like YouTube, allows users to upload, view, rate, and share videos for free, including professionally produced content. One hundred million visitors per day (the combined populations of Canada, Poland, and Australia) engage in fifty-seven thousand searches per minute of videos that are divided into categories such as oral, anal, blond, ebony, MILF, cuckold, squirting, teen, and *My Little Pony*. The site is owned by Montreal-based

MindGeek, which also controls the other major adult "tubes": YouPorn and Red Porn, as well as the production studios Brazzers, Digital Playground, Reality Kings, Sean Cody, and Men.com (the latter two of which feature gay and bisexual performers). MindGeek's virtual monopoly in the industry, along with abundant free X-rated content (despite "no nudity" policies) on sites such as Tumblr, Instagram, Twitter, and Snapchat, has made porn simultaneously ubiquitous and, for its performers and producers, largely unprofitable—not unlike the effect of Napster and YouTube on musicians. In the process, it's also given minors historically unprecedented access to sexually explicit media, most of it reflecting traditional gender dynamics and shot from the male's point of view—often a hostile, narcissistic male at that: sex is portrayed as something men do *to* rather than *with* women. Women exist primarily as providers of male orgasm (the goal of every encounter); their own pleasure is presented as a performance to that end. The men themselves appear joyless, mechanical. Given its boundless novelty and availability, contemporary porn raises questions about how young people's erotic imaginations will be shaped long before they've engaged in so much as a good-night kiss: the impact it may have on desire, arousal, behavior, sexual ethics, and their understanding of gender, race, and power in sexual expression.

Absent other sources, it's clear that teenagers turn to porn, at least in part, for sex education, even as they claim to know that its content is about as authentic as pro wrestling. Girls in particular consult sexually explicit media for a template of male partners' expectations, despite being more than twice as likely as boys to be disturbed by its treatment of women. One national survey of teenagers and their parents found that the teens watched significantly more porn, and harder-core porn, than the adults. While more than 56 percent of the boys and 38 percent of the girls had seen pornography (a figure that skews low because it includes very young teens), only a third of their fathers and fewer than a fifth of their mothers had. The boys were at least three times more likely than their dads to have watched videos depicting facial ejaculation, double penetration, BDSM, coercive sex, gang bangs, and rape; the differential between girls and their mothers was generally even higher.

I never asked a boy in my interviews *whether* he watched porn (nor, unlike with girls I interviewed, did I ask *whether* he had ever masturbated). That would have shot my credibility to hell. Because of course all of them—every single one—had. Instead, my question was when they first saw it. Often, the first exposure was unbidden: older brothers (or older brothers' friends) spun around a smartphone screen or motioned them over to a computer, either as a manly rite of passage—doing the little guy a favor—or to freak them out. Or maybe a friend sent a link to a video of a naked woman masturbating or a clip of a man shoving his outsize penis down a woman's throat. Boys typically considered such pranks to be, yes, "hilarious."

Those involuntary sightings tended to happen younger, at age nine or ten. Intentional searches started, for most, somewhere between sixth and eighth grade—usually before the boys had masturbated or were able to ejaculate. Initially, they'd type something rudimentary into Google like "boobies" or "naked women" (though what is returned by those searches is hardly innocuous); they often described their maiden glimpse of female genitalia as more alarming than arousing. "When I was in sixth grade, boobs were really cool," said a high school senior in San Francisco. "But I couldn't handle vaginas. They just look kinda weird, like an alien mouth." What was originally disturbing soon became compelling. Over time—and here's where this generation is unique—most of the boys learned masturbation entirely in tandem with porn, yoking it to their cycle of desire, arousal, and release. "I started masturbating in sixth grade, about a year after I was first shown porn," said Mitchell, a sophomore at a Los Angeles college. "I don't think I masturbated without it until at least tenth grade. It was just so easy to get, I didn't consider *not* using it. You go on Pornhub and there's all these categories you can go through. And being able to reach the normal stuff and the weird stuff equally easily was crazy." Cole, the guy from the previous chapter who is now attending a military academy, recalled, "I have a friend who was a legend among the high school crew team. He claimed that he'd stopped using porn completely. He said, 'I just close my eyes and use my imagination.' We were like, 'Whoa! How does he do that?' "

With such infinite variety, boys didn't limit themselves to just one or two videos at a time: "What I do is, I have three or four favorite sites," explained a junior at an East Coast college. "I look at all the thumbnails on the first couple of pages to see what seems good. Then I open the tabs for those videos on the first site, then the second site, and so on. Then I go back and forth between them, skimming through until I find something that really gets me off." Using porn was so effortless, so reflexive, that, sometimes, he found himself gravitating to it when he didn't intend to. "My fingers will just start clicking and I'll be like, 'Wait, no! I meant to go on CNN! I didn't want to go on Pornhub! But it doesn't matter. I do it whenever I'm bored, depressed, stressed out—whenever I need something to do."

Although all the boys I met had watched porn—one equated his morning session to sneezing—they didn't all relate to it in the same way. A small group, after some experimentation, had rejected explicit media entirely. "In middle school, porn was considered cool," a senior at a Southern California high school told me. "Guys knew the names of porn stars. And I watched it almost because it was like the unknown—like the same impulse that makes me want to climb a rock or go to a forest. But pretty quickly, I was like, 'This is just so fucked up.' What's on the screen isn't actually *sensual*, not for either person. And often the only part that's touching between the two people is their genitals. That's literally all that's touching. And I was like, 'What am I watching?' It doesn't make sense to me, how you can look at a woman with tremendous respect and then go watch porn. So I stopped."

A second group felt that their porn use had no effect on them, many of them asserting, "I can tell the difference between fantasy and reality." That, as it happens, is the instinctive response people give to any suggestion of media influence. None of us wants to think we're so impressionable, though we're quick to recognize that others are (several of the boys I interviewed made grim predictions about the impact of iPhones, video games, social media, and porn on "the next generation," by which they meant their slightly younger siblings). But decades of research show that what we consume becomes part of our psyches, unconsciously affecting how we feel, think, and behave. When false information is embedded into a fictional story, people will come to believe it (yes, reader, you would, too), and those beliefs are strengthened over time. Consider the college students who were given a short story containing the bogus "fact" that exercise weakens the heart and lungs. When questioned directly afterward, they were unconvinced this was true, but two weeks later, they had become certain it was. Karen Dill-Shackleford, a media psychologist at Fielding Graduate University, speculates that there is something about the way we suspend disbelief when swept up in a story that opens us to uncritically and even permanently accepting its reality (that theory also partly explains both the spread of fake news and why the repetition of obvious lies can be a successful ploy for political candidates).

Still, assessing porn's impact isn't easy. For one thing, as with many issues related to teens' sexual health, the subject is not exactly a fount of research funds; designing an ethical study, especially given that minors aren't supposed to be watching in the first place, would be a challenge. What's more, disentangling porn's influence from that of family, peers, school, and mainstream media is tricky. Much of what is known is more correlative than causal. Even so, the findings are noteworthy. On the upside, there's some indication that porn has a liberalizing effect: heterosexual male users, for instance, are more likely than their peers to approve of same-sex marriage. On the other hand, they're also less likely to support affirmative action for women and to disagree with or only tepidly endorse the statement "I want there to be equal numbers of men and women who are leaders in work, politics, and life." And it only goes downhill from there. College students who regularly consume porn (which some researchers define as as infrequently as once a month, a bar so low that it made boys I spoke to either blanch or burst out laughing) are *more* likely than others to consider its portrayal of sex to be realistic. Porn use has been associated with boys' real-life badgering of girls for nude pictures. Both boys and girls who consume porn at younger ages are more likely to become sexually active sooner than peers (although greater media use of any kind—TV, internet, music, movies, video games—is associated with precocious sexuality regardless of race, gender, or parents' education), to have more partners, to have higher rates of pregnancy, to view sexual aggression more positively and women more negatively, and to engage in the riskier and more atypical behaviors porn depicts. Heterosexual anal sex, for example, has historically been vastly more common in porn than in reality (and usually shown as easy, clean, safe, and, perhaps after some initial protest, pleasurable for women), but that ratio is starting to shift. In 1992, 16 percent of eighteen- to twenty-four-year-old women in the US reported trying it; by 2009, that percentage had risen to forty; 20 percent of eighteen- to nineteen-year-old girls now report having engaged in anal sex and about one in seventeen fourteen- to seventeen-year-olds (most, by the way, do not use condoms).

I'm not trying to demonize a particular behavior, but it's important to look at young people's motivations as well as the quality of their experience. A 2014 survey of anal sex among heterosexuals ages sixteen to eighteen found that for boys, porn was a driving factor. They viewed the act not as something to share with a partner, whom they believed would need to be (and could be) coerced into it, but as a form of competition with one another. Their strategies included a "try it and see" approach: inserting their finger or penis into a girl's rectum without her consent and hoping she'd acquiesce. (*The Mindy Project* did an entire episode on this, in which the star's boyfriend claimed, "It slipped.") One boy said he would simply keep at it until he wore a girl down and she'd "get fed up and let you do it anyway."

Another finding: among heterosexual women, including those who currently watch the same levels of porn, sexual submissiveness and compliance were correlated with the age at which they first started viewing. According to Paul Wright, an associate professor of media at Indiana University who conducted that research, "It suggests that if I'm female and I see porn at a very young age, I see, *This is what sex is.* So later on, I'm going to initiate that behavior because I think that's what a guy wants. Or if he wants to do something, I'll be more likely to just go along with it. If I start viewing when I'm older, I'm more like, 'Um, this is what people do in porn, not what they do in real life.' So if a guy is like, 'Hey! Can I strangle you and come on your face?' they're more likely to say, 'Go to hell.' That's our explanation, anyway. We don't know for sure."

Although Wright has yet to analyze his data on men, he hypothesizes a similar association between age of first porn use and expressions of sexual dominance. His previous coauthored research found associations both in the US and internationally between porn consumption and men's desire to engage in a range of aggressive behaviors, especially if the men drank regularly before sex. Among male college students, porn use has been correlated with seeing sex as a purely physical experience; regarding girls as "playthings," objects of conquest rather than people; and measuring virility, social status, and self-worth by the ability to score with "hot" women. College students of both sexes who report recent porn consumption have repeatedly been found to be more likely than others to believe "rape myths": that only strangers commit sexual assault or that the victim "asked for it" by drinking too much or wearing "slutty" clothing or by going out alone.

What may be of more immediate concern to guys themselves, though, is that male porn users report *less* satisfaction than others with their sex lives, their own performance in bed, and their female partners' bodies—and the effect becomes apparent among those who indulge as rarely as a few times a year. There is even speculation that because of its convenience as well as low physical and emotional investment—porn never rejects you, never makes demands of you, never wants you to talk about your feelings—the rise in porn use is partially responsible for the lower rates of intercourse among millennials: not, I imagine, quite the victory abstinence-only advocates were going for. That reduction of pleasure in partnered sex was what concerned the majority of my interviewees. Even when they felt their porn habit was at reasonable levels, more than half had, at one point or another, cut back on their use, much the way they would if they were drinking too much or smoking too much weed. A high school senior in New England told me he took a break when he found himself daydreaming during math class about a female friend who sat across from him, and realized that he wasn't fantasizing about what she'd look like naked, or even what it might be like to have sex with her. "I was thinking about what she'd look like with cum on her face," he said. "That was a wake-up call for me."

Reza, a sophomore at a Boston college, believed porn increased his awareness of real women's physical imperfections. "I've got things narrowed down to a very, very specific body type that turns me on," he explained. "Like the size of the areola and its color, that sort of thing. It's probably not all driven by porn, but I figured out what I liked from that and I think I wouldn't have otherwise. It doesn't ruin my relationships, but it's not nice when I'm trying to talk my girlfriend into liking a part of her body, but I'm secretly thinking, *Well, actually, I would prefer . . .*" And Kevin, a high school junior from Kentucky, said that after masturbating to "all those skinny white women" (he's Caucasian), he was having a hard time becoming aroused by his African American girlfriend's body, which was neither of those things. "I still watch porn five or more times a week, though," he admitted. "It's so easy, and I'm a lusty, young teenager. But sometimes I wonder if I'm only depriving myself of a better sexual relationship and that's kind of depressing."

Some boys fretted more over their own bodies' contours than their partners', especially (and perhaps not surprisingly) their penis size. That has always been true, to a degree, but porn has exacerbated the issue. "Everyone watches porn and then gets super nervous about their dick size," confided a college sophomore in Chicago. "I mean, it's brutal. Like if you're in the locker room, you're going to turn around and try to hide yourself, or you're not going to change in front of other guys." Some responded by shaving their genital area: as with girls, porn was creating new expectations around hair removal, though for different reasons. Girls generally cited a concern (however inaccurate) about hygiene along with a fear of humiliation by male partners; boys thought it would make their penises look larger. The results weren't always what they'd hoped. "No one told me that if I did it, I'd look like a baby," one high school senior said. "I looked down and it was like I hadn't gone through puberty yet!"

A few boys were so concerned about size that they avoided sexual situations. "I had a girlfriend in tenth grade," Mitchell, in Los Angeles, said, "and as we started being more sexual I became very nervous about the idea of getting erect and about being . . . *sufficient*. I couldn't get hard during our first real sexual experience because that was so much on my mind. And once you feel like you can't get it up, you can't. You're done." For a year, Mitchell declined his girlfriend's offers to perform oral sex on him, afraid that if she saw his penis, she'd be disappointed. "She was like, 'What kind of guy wouldn't want head?'" he recalled. "It was hurtful." With time, and maybe a little maturity, he got past it. In retrospect, he said, "Comparing myself to porn was obviously ridiculous. But, you know, it's also kind of understandable."

Like every boy I spoke with, Mitchell claimed to know that, *of course*, porn wasn't realistic. But that line between fact and fiction was less clear; after all, porn is depicting *something*, and what other point of reference do young people have? "If you're a teenage guy and you don't have much sexual experience, and you've been watching porn or watching HBO at two a.m. for the past six or seven years, you can develop almost a . . . *fear*, really," said a college sophomore in North Carolina. "A fear that you would not be able to perform up to those standards, though, of course, no one really can. But maybe the starkest contrast is your perception of the kind of feedback that you're going to be getting from a girl. Like that they will be moaning and having orgasms all over the place. That's obviously not the case."

For a small but significant group of the boys I met—one that included Mason, the guy with the flip phone—porn use had become a compulsion, one they felt had seriously harmed them. Some had gone deep into the world of fetish videos, unearthing clips of, say, men eating women for erotic pleasure, or the reverse (I am not referring to oral sex here, but cannibalism). Some had, at least for a time, joined the "no fap" movement ("fapping" being vernacular for masturbation), becoming "fapstronauts" who forswore self-love entirely in order to "retrain" their brains. I can't say whether porn obsession itself was the primary issue for these boys or a preexisting personality trait that found its expression there. What was important, though, and consistent, was that they believed porn had been damaging in ways that no adult had ever discussed with them, and that they had never previously discussed with an adult.

The New Sex Ed

Mason's parents never talked to him about sex, sexual ethics, or healthy relationships; that was true of most of the boys I met, but some felt their parents had nonetheless led by example, through the loving, respectful, even playful ways they interacted with each other. That was not the case for Mason. His parents hadn't slept in the same room since he was in elementary school. At one point, they didn't speak for an entire year. His father mocked his mother, spoke over her when she talked, ridiculed her opinions. Mason encouraged her to divorce him, but as a devout Catholic, she refused.

Meanwhile, as puberty loomed, Mason became curious. His teachers repeated abstinence-only platitudes; his parents were silent. Inevitably, then, he turned to the modern-day street corner: the internet. In seventh grade, he Googled "Playboy" (which one guy I spoke with referred to as "What's it called, you know, that famous porn magazine with photos and stuff?"), a search that, when I tried it, instantly returned dozens of images of partially nude, fully acquiescent women: a minor reality TV celebrity dressed in sneakers and a white bikini bottom, her tank top pulled down to expose one nipple; a big-bootied woman squatting in a thong and "fuck me" pumps, shot from behind, her face turned coyly to the viewer; a woman straddling a rear-facing chair, her legs stretched wide, wearing only tuxedo cuffs, a bow tie, and bunny ears; another in lacy panties, crawling toward the camera on all fours, licking one raised hand as if it were a cat's paw. All of them with identically dyed blond, flowing manes; flawless, oiled skin; ample breasts; full bottoms. All of them seeming to make direct eye contact with the viewer, pouting invitingly. At the same time, though, the photos were only marginally more explicit than an average music video, and some not even that. Next, Mason tried making an end run around his parents' safe search filters by slightly misspelling terms: say, "big bobbs." That got him further. He had his first orgasm at fourteen, watching a video of a naked woman trickling water onto her breasts. And that, he said, was when "the whole spiral started. At first it was simple things like a girl in a pool. Then it was people having sex. Lesbian stuff. Anything in the 'Big Tits' category. Anal stuff. Double penetration. I'm sure I've seen a few things involving every fetish category. I've seen some crazy shit."

One night, Mason was sitting on the basement couch, browsing porn on his school-supplied iPad, when his dad walked in.

"You shouldn't be watching that," his father told him. "It's bad for you."

As it happened, Mason had recently snooped on the older man's laptop and found a trove of bookmarked porn. So he snapped back: "Don't be a hypocrite. I've seen all the stuff *you* watch."

Without another word, his father plopped onto the couch and turned on an episode of *Bob's Burgers.* When it was over, he went upstairs to bed. End of discussion. "I feel he sort of failed me," Mason said. "I kind of wish . . . Even after I fired back at him, there was still an opportunity for him to have that conversation with me. Maybe if he had said, 'This will skew the way you view women. It's not real. And it's not going to help you get a girl; it's only going to keep you from interacting with girls in a healthy manner,' that might have made a difference for me. But my parents were too fearful to actually deal with any of it."

By his sophomore year of high school, Mason was spending about an hour a day immersed in porn. He watched in his bedroom. He watched in the den. He watched in the bathroom stalls at school. He watched at parties. It was, he said, as integral to his life as brushing his teeth, eating, drinking, sleeping. "You can spend so much time in this porn fantasy world that you don't even look up and live your real life," he said. Once, he popped in his earbuds and scrolled through his downloads in the back of the family car after a dentist appointment (no drilling jokes, please), as his brother and mom chatted in the front seat. "That's how normalized it became," he said. "If it was going to be thirty minutes in the car, I might as well watch porn."

Although at this point Mason had never kissed a girl or even held hands, he said that he half expected that at any moment a "hot woman" would appear out of nowhere and demand to have sex with him. "That was my whole perception of how it was supposed to go," he said. "In porn, women are portrayed as these sexually driven animals. It's all they want; it's all they care about. It warped my perception for so long." He didn't realize how much, though, until, at age sixteen, while surfing a Russian social network that allowed him to circumvent his parents' safety filters, he watched a video of a woman defecating into a hotdog bun. She poured condiments on the result and handed it to a second woman, who ate it.

I asked Mason if he found that arousing.

"It was, kind of," he admitted. "Outside of that specific context, I would find this obviously disgusting act to be repulsive. I *do* find it repulsive. But it was being portrayed as extremely sexual. That's what porn does. Just media in general. Sexualizes anything. Sexualizes people eating poop. And it boggles my mind how easy it was to be drawn in by it all."

ALTHOUGH MASON VIEWED more porn and more extreme content than most of the boys I talked to, he was hardly alone. "I don't consider the porn I watch to be representative of the person I am," said Daniel, a freshman at an elite Midwestern college. Daniel had a lantern jaw and hipster glasses; he wore a backward baseball cap that he continually adjusted, running a hand over his curly hair. "Like, the whole category of 'Unwilling' [women who say no to sex, then change their mind when 'force fucked']. It's very appealing to me even though I know it's wrong. And I do truly believe it's wrong. I would never do it. But I'd be lying if I said I didn't enjoy watching it." Daniel was from Michigan and grew up in a very different family from Mason's: his parents were happily married, and they identified, at least nominally, as Jewish. His mom, an instructor at a community college, was a feminist. In real life, Daniel was consciously trying to curb his use of the thoughtlessly sexist, homophobic language that had been common in his high school. He also said he considered any form of sexual interaction to have "spiritual significance" and claimed to prize intimacy over "raw sex."

But that's not what got him off. He'd had three hookups in college; while he could get an erection, he wasn't able to orgasm in any of them. He'd also found it "a bit of a struggle" to climax during intercourse with his high school girlfriend, with whom he'd been in love. Their sex wasn't stimulating enough; it wasn't intense enough—it just wasn't *enough.* "I felt like I was never really satisfied," he said. "There was always more to try. Like, 'Oh, this is pretty good, she's letting me do a lot, but we haven't done *this* yet, we haven't done *this,* done *this,* done *this* . . .'" The girl herself, who was a year younger than him, also seemed to have taken her cues from porn. She would writhe and moan when Daniel jackhammered a finger into her vagina. When he told her he wanted to "take a break" from their relationship, she offered up anal sex to change his mind (although that idea could equally have been inspired by mainstream films such as *Kingsman,* in which the hero's incentive for saving a Scandinavian princess is her promise that "we can do it in the asshole"). He accepted, but the experience was nothing like what he'd seen on-screen. "It was really difficult," he recalled, "and it hurt her intensely. She was in pain. That was nothing I wanted to see. It was fucked up. And I felt like shit afterward."

The two did eventually split up, briefly; they reunited after a round of rough sex in an elementary school parking lot, during which he choked her, pulled her hair, and, at her request, slapped her across the breasts. That was the first time Daniel climaxed without using his hand. "It was the best sex we ever had," he recalled. Certainly, people have predilections, but it's troubling that both instances of aggressive sex Daniel recounted happened when his girlfriend believed their relationship was in jeopardy. Daniel himself considers his behavior to be contrary to the intimacy he claims to crave. He doesn't know whether that disconnect is natural, part of his wiring as a sexual person, or shaped by his repeated consumption of prepackaged fantasies.

As another boy, a high school senior in San Francisco, put it, "I think porn affects your ability to be innocent in a sexual relationship. The whole idea of exploring sex without any preconceived ideas of what it is, you know? That natural organic process has just been fucked by porn."

We tend to believe that genital response—becoming erect for men, lubricating for women—is synonymous with arousal, but that's not always true. Those are automatic physiological reactions. Emily Nagoski, a professor of health behavior who studies the science of desire, likens it to being tickled by someone who infuriates you: your laughter is involuntary; it doesn't signal enjoyment. For men, the overlap between blood flow to the genitals and "turned-on" feelings is only 50 percent—which sounds low, until you hear that for women it's a mere 10 percent. The fancy, scientific term for this is "non-concordance" and it's why a woman can, say, orgasm during a disliked sexual activity, including rape: despite what *Fifty Shades of Grey* (or some politicians) would have you believe, that does not make her "in denial" of her secret wishes. Similarly, a man may be sickened by an image that makes his penis rise; that isn't the biological version of a poker tell. Bodies react to what is perceived as sexually *relevant*, Nagoski said, not necessarily to what's sexually *wanted*, appealing, or enjoyable (you can also experience the inverse: finding something sexy without having a genital reaction). When a boy gets an erection in response to scatological porn or a report of rape in his campus newspaper, then, it doesn't necessarily mean he's into it; his body may just be acknowledging something involving sex is afoot.

That said, when something is both sexual and taboo, such as eroticized violence or extreme degradation, it can trigger a mental process similar to the one that makes you obsessed with polar bears as soon as you're told not to think about them. "Embedded in the idea 'Don't respond to that sexually relevant information,'" Nagoski explained, "is the idea '*Respond* to that sexually relevant information.'" The tension of that contradiction can turbocharge arousal, making something objectionable *more* exciting than that which is overtly desired. For adolescents, whose brains (not to mention hearts) are still developing and acutely influenced by early exposure, the impact can be significant. While porn isn't "addictive" in any classical sense, their sexual response can be conditioned, trained to be narrowly and exclusively triggered by its images—even when (or perhaps especially when), like Daniel, they find those images offensive. The answer is not to stop masturbating, Nagoski emphasized, but to do so in a much wider variety of contexts—using one's imagination, reading sexy books (she suggests romance novels, which have become both more explicit and more realistic about female pleasure) with a romantic partner in the room, using techniques beyond a porn-induced "death grip" on the penis—and learning to differentiate between what is highly arousing and what is actually enjoyable.

Most of the boys I talked to had instinctively backed away from the web's more gonzo options, especially any they felt were "rapey." Even so, disturbing content—or what would once have passed for disturbing content—can break through. After one boy told me that he liked to look up blow jobs on Tumblr, I checked it out, and somehow ended up with a GIF of a woman fellating a dog. That brief exposure is something I can never unsee, and, truly, I felt it degraded my humanity. "I'd say my exposure to porn has desensitized me to normal or vanilla sex," a high school senior in San Francisco told me. "But that's also a function of the access my generation has to the internet in general. Having spent enough time and seen enough gore, I have become utterly indifferent to even the most graphic sex and violence. Nothing in porn is shocking or weird to me. It's entertainment, like an action movie."

In an oversaturated media marketplace, attention is the most prized currency. The best way for porn to capture and keep its share is to perpetually up the ante on aggressive and cruel acts—face fucking! Bukkake! Stealth "creampie"!—none of which are likely to result in orgasm for most women. Quite the opposite: some recent porn trends are so physically demanding that female performers later require surgery to repair prolapsed vaginal walls or anuses (there is little regard for women's safety on set, and certainly no worker's comp or disability pay). Porn is also pushed to extremes by conventional media, which has become ever more explicit in its own quest to lure viewers. Even so, most heterosexual porn still centers on "the big six": digital stimulation of the vagina by a partner, self-stimulation (by either sex), fellatio, cunnilingus, and coitus. Those acts tend to appear together, according to Bryant Paul, an associate professor of media psychology at Indiana University who studies the impact of porn. So if you see one in a clip, you're likely to see them all. But "the big six" also often cluster with certain, harsher acts: men spitting on women, "facial abuse" oral sex, anal sex. What's more, specific behaviors don't tell the whole story. Pornhub's front page features the most popular videos in a viewer's home country; in the United States (the world's biggest consumer of adult content) that means an array of pseudo-incest porn (stepmothers, step-siblings, teenage "besties" who have sex with each other's "dads"). Even if the sex itself is fairly routine, the taboo that it flirts with to amp up arousal—the fantasy it promotes, reinforces, and possibly normalizes—is not.

Paul has also found that, depending on the context and intensity, young adults don't always register aggression in porn. They'll notice if it's integral to a particular scenario, such as BDSM, or a woman being choked while crying; but when the aggression is "light" during high-intensity sex—slapping, hitting, choking, hair-pulling, name-calling—and less central to the action, they retain no memory of it. "We had them rate various acts before they watched the clips, so we knew they perceived them as aggressive," he said. "But when we'd say, 'Name all the aggressive behaviors you just saw,' they would say, 'I didn't see any.' So, we tried asking, 'Did you see any of the following acts?' and listing them, but they didn't even get it when we put it that way." What could that mean? "One possibility is they literally aren't noticing because it's not what they're there to see," Paul suggested. A more likely explanation, though, involves something called "priming theory": how media of any type unconsciously promote certain associations in the mind. "So if you're noticing these things implicitly," Paul said, "but are disregarding them because it's not what interests you, it's not what you're there to see, that could lead to the 'prime' of 'ignore microaggressions toward women.'" Or, potentially, to "ignore not-so-micro aggressions": it may, for example, partly explain why men who use porn regularly, as mentioned earlier, are less favorably inclined toward affirmative action for women or toward equality in political and professional life.

It's Not the Sex; It's the Sexism

Obviously, media "primes" about gender and sexuality don't come exclusively from porn. They're found in video games, advertising, movies, TV, music videos, song lyrics. "The entire Adult Swim lineup [on the Cartoon Network] probably has more potential to influence people's beliefs about sex than porn," Bryant Paul commented. "Because it's mainstream, which makes it seem more acceptable. It's kind of like how twenty years ago, if you read something in *Hustler*, you'd know it was fantasy, but if you read the same thing in *Maxim* you might try to incorporate it into your bag of tricks." So, while an episode of *Family Guy* in which bikini-clad Asian women are accidentally released from a character's garage and car trunk (*hilarious!*) may not make viewers believe it's okay to take Asian women hostage, it may well contribute to more readily treating them as objects than white women. Although porn has become a lightning rod for concern, then, it may really be more a symptom than the disease itself. Nor is its sexual explicitness an inherent problem: it's the persistent depiction of women as things, and in degrading and compliant roles—less the *sex* than the *sexism*. And that is abundantly on display in the mainstream media boys consume throughout their lives.

There is broad agreement that the incessant sexualization of women in media hurts girls. Even a brief exposure—to two thirty-second advertising spots embedded in a four-minute reel—has been found to undermine body image, erode self-esteem, and trigger self-objectification. Yet, rarely do we discuss how those one-dimensional images of women influence *boys'* perceptions of female peers: the truth is, it affects them profoundly. One clue as to how can be found in a 2000 study by researchers from the University of Massachusetts. They edited together scenes from the R-rated movies *Showgirls* and *9½ Weeks* that were judged as degrading to women, depicting them as objects to be exploited or manipulated sexually. None contained violence—they included scenes of a striptease, of a blindfolded woman—but they emphasized male dominance as well as female submission and availability. They also portrayed male, but not female, sexual satisfaction. Half the college-age participants watched those clips, and the other half, the control group, watched a cartoon from an animation festival. Afterward, both groups read a magazine account of either acquaintance or stranger rape. While there was little difference in their response to the stranger scenario, the men who watched the degrading videos were more than twice as likely as the control group to agree with the statement that the victim enjoyed the acquaintance rape and secretly "got what she wanted." The effect held steady regardless of the men's attitudes toward gender roles or sexually explicit material. Such sexualized images, again and again, appear to "spill over" and affect men's perceptions of all women. In another experiment, men who had been shown objectified images of women were asked to rate the competence and intelligence of a female researcher for an unrelated task; they gave her lower marks than men who'd been shown neutral images.

From the earliest ages, children are subject to messages that present women primarily as objects for male use, as rewards for victory, wealth, and fame; messages that disregard women's perspective and inaccurately represent their gratification. Parents of little girls may surround their daughters with books and movies and images of complex female characters in an effort to offset all that, but rarely do parents of boys do the same: the distorted depiction of women in media is seen as a problem for girls alone. Yet little boys watch professional sports, where the sole women on the field are cheerleaders. Female characters in family-friendly, G-rated films are depicted in revealing clothing at about the same rate as in R-rated films. That, of course, is when women or girls are present at all. Despite their rising visibility in franchises such as *The Twilight Saga*, *The Hunger Games*, and *Star Wars*, by 2018 only 31 percent of top-grossing Hollywood films featured female protagonists—a record high!—and women comprised just 35 percent of all speaking roles, a figure that has held steady over the ten years researchers have tracked it.

By their teens and twenties, boys consume content from men's lifestyle brands whose presentation of female sexuality is not always so very different from porn's: study participants presented with quotes from popular "lad" magazines mixed with those by convicted rapists were unable to reliably distinguish between the two; they actually ranked the former as more derogatory. "[Such publications] give the appearance that sexism is acceptable and normal," the researchers concluded, adding that "the legitimization strategies that rapists deploy when they talk about women are more familiar to these young men than we had anticipated."

Rape is routinely used as a plot device on premium cable programs that draw a huge teen audience, such as *Game of Thrones* or *Westworld*, where it is presented as a presumed male fantasy. Movies and TV shows such as *Fight Club*, *The Godfather*, *The Wolf of Wall Street*, *Empire*, *Billions*, and *Silicon Valley*—all iconic among high school and college guys—consistently define masculinity through wealth and dominance; the worst (and often cleverest) slurs against other men invoke femininity, homosexuality, sexual submission, or rape. Weak men are "pussies"; a person beneath contempt is "fucked" or in the position to "suck my dick." Aside from the occasional tough bitch who trumpets the size of her own balls (the MBA version of Gillian Flynn's "cool girl"), women are invisible, love interests, sex workers, or trophies through which men assert power over other men. Nor is it generally clear whether the point is to critique that culture or to celebrate it. Women in *The Wolf of Wall Street*—prostitutes, strippers, generic hot chicks from central casting—are portrayed as the rightful spoils of success (and excess). Until its creators were shamed by the internet, *Silicon Valley*'s cast included precisely one woman: the only conventionally "hot" character on the show, she was an assistant (neither a programmer nor even an executive) who played Wendy to a start-up's lost boys. The show's continual stream of phallic humor, while indisputably funny, cements as much as satirizes brogrammer culture. *Billions*, too, reinforces the behavior it purports to expose. Misogynist barbs, pornification, and the marginalization or objectification of female characters are all so pervasive that it's hard to identify just one (or two or three or five) examples. Though one can try: there is the character who is nicknamed "Pouch" because "he's got no balls. They say he has a vagina." Or the one who yells, "You wouldn't know the right thing to do if it kneeled down and sucked your tiny goddamn cock!"

Wrong. Wrong. Wrong. Boys *know* such sentiments are wrong—they are not completely blank slates for the culture to inscribe. Still, they are barraged by these images and ideas, usually without challenge or context. About half of boys ages ten to nineteen say that at least several times a week on TV, in movies, in music videos, or on YouTube they see unrealistic images of women's bodies or women whose bodies and looks are portrayed as more important than their brains or abilities (and, frankly, a mere *half* of boys seems low). Nearly half also see female video game characters portrayed as "hot" and a quarter see such characters every day. "I've been watching the show *Californication* lately," said Mason, the guy with the flip phone. "I think in some ways it's more damaging than things that are incredibly unrealistic. Because it's just *slightly* unrealistic. It's still kind of believable. Like, the main character has sex with everyone wherever he goes. His character has such a good build, it's believable. But every girl in the show is a topless, sex-crazed fiend, and every dude is sex-crazed as well. And, I don't know. They make it seem so *convincing*. Whereas if you were to watch a porn video where a dude comes in with his dick in a pizza box, it's like, 'All right, obviously that isn't going to happen in real life.' Also *Californication*—it's on regular TV; it's not porn. So you believe that more." Even a show like *Modern Family,* Mason added, which was one of his favorites in high school, includes Gloria, the dumbed-down, hypersexualized (though secretly smart!) bombshell who is essentially a latter-day Charo. "Whatever media you turn to," he said, "however family friendly it is, it's always sexualizing women. It's inescapable."

Remember, media scripts influence real-life emotions and behavior, *even when we think they don't.* And the media scripts boys consume from childhood onward are continuously objectifying, demeaning, hostile, inimical, or indifferent to women and present masculinity as inherently antagonistic toward femininity. No wonder that, in 2017, the first woman credited as writer on an episode of the animated Adult Swim hit *Rick and Morty* was inundated by rape and death threats from the show's male fans (its creator, Dan Harmon, himself later acknowledged that he'd sexually harassed and attempted to destroy the career of a female underling with whom he worked on a previous series). As many as two-thirds of female journalists (this one included) have also been harassed online, receiving threats of rape and murder or nasty comments about their appearance. Female gamers and Twitch streamers are subject to a torrent of misogynist vitriol. Video games, incidentally, are now a $140 billion annual global business, as opposed to Hollywood's mere $38 billion; players' exposure to sexual objectification and violence against women within them has been associated (it begins to feel tedious to report it *yet again*) with higher rates of sexism, "rape-supportive" attitudes among men, and tolerance of sexual harassment. Playing sexually charged video games for as little as twenty-five minutes has been correlated with an increased tendency among college men to say they would engage in "inappropriate sexual advances"—that's particularly striking, given that, in a laboratory setting, boys would presumably want to portray themselves to researchers as more principled than they might, in fact, be.

Then, of course, there is hip-hop, which in 2017 became, for the first time, the most popular music genre in the US and attracts the youngest audience. Like porn, video games, and mainstream media, the category is both broad and deep, encompassing Kendrick Lamar as well as Migos; Dope Saint Jude and Nicki Minaj. Historically a crucible of creative social protest, in its most commercially successful, corporate form, it earns billions annually by distorting black masculinity, exploiting black women's bodies, and degrading female sexuality. "I think music has some of the biggest impact on how guys treat girls," a high school senior in the Bay Area told me. "In the car, my friends and I listen to all this stuff that's just 'fuck that bitch and then quit her' over and over. When you hear that, like, five, six, ten times a day, it makes it hard to escape having that mind-set." Tricia Rose, director of the Center for the Study of Race and Ethnicity in America at Brown University, refers to the "gangsta-pimp-ho trinity" whose most devoted fans are typically white, suburban boys who may never have personally known an African American. She has written that although hip-hop didn't invent sexism, nor is it the first musical style to profit from it, its power, imagination, and fusion of materialism with strip-club culture has made misogyny "sexy, visible and funky. . . . And, frankly, if you want to find openly celebrated sexism against black women, there is no richer contemporary source than commercial, mainstream hip-hop."

Some male artists, as they've aged, have expressed passing regret for their earlier attitudes. In January 2018, Jay-Z spoke out in support of the #MeToo movement, despite having used the word "bitch" in 109 of his 217 songs (at that time), building his early brand with lines like "I'm a pimp in every sense of the word, bitch." Nas, who once penned lyrics about urinating "on bitches who famous" and "putting stitches in they anus" lamented in his song "Daughters" that "God" gets revenge on the "coolest playas and foulest heartbreakers" by making them fathers of "precious little girls." And on his 2018 single "Violent Crimes," Kanye West described "niggas" as "savage" "monsters," "pimps," and "players" until they have daughters. Then, "Father forgive me," he wrote, also appealing to a higher power, "I'm scared of the karma." All three men changed their literal tunes after becoming dads. Certainly, any positive change is commendable, and I'm glad they've seen some hazy form of the light, but as culture critic Dara Mathis tweeted about West, "Daughters are not spiritual retribution for your misogyny." They are neither a form of punishment (among other things, what would that make sons?) nor "tiny spiritual guides sent to newly show you the humanity of girls and women." Nor did Ye's hand-wringing ultimately amount to much. Six months after "Violent Crimes" debuted, he dropped "I Love It," with Lil Pump. In the video, as they repeat "You're such a fuckin' ho, I love it," the two men shuffle down a hall lined with alcoves containing naked, faceless women who have been forced to their knees, their hands behind their backs, presumably bound. The only real difference between that—which was the biggest-ever global hip-hop debut on YouTube—and actual porn is that Kanye's video was produced by the mainstream Warner Brothers and performed, albeit with more TV-friendly lyrics, on *Saturday Night Live.*

The Body Is a Vulnerable Thing

"I'm not going to lie," Mason said. "I feel kind of like I'm playing catch-up when it comes to where other people are at in terms of their sexual experiences."

Porn did not accelerate Mason's real-life sexual pursuits. Quite the opposite. He even avoided the simple act of kissing for much of high school, afraid that without "experience" he would do it wrong. Nothing he'd seen on screen prepared him for how awkward real-life encounters could be, or how to navigate them with a partner. He thought he was just supposed to *know.* Mason's high school girlfriend performed oral sex on him for the first time when they were seniors largely, he suspects, out of guilt for refusing intercourse (her "no" as he prepared to enter her was so soft he almost didn't hear it). Since they never talked about it, though, he can't say for sure. At any rate, he lost his erection during the act, so she stopped and they got dressed in silence.

So, I asked Mason, when did you have intercourse for the first time?

He gazed up at the ceiling, fiddled with the pompon on his hat, squinted as if he were trying to recall. Finally, he said, "Friday."

"You mean, five days ago?" I replied.

He nodded.

He'd met the girl, Jeannie, just a few weeks before at a party; they'd both been so high on a strain of weed called "God's Pussy" that they couldn't get off the couch. Once they'd rallied, she invited him back to her dorm and they started fooling around. When she straddled him in preparation for intercourse, though, he went soft. He blamed the weed—though it's not actually clear whether marijuana affects erections the way alcohol can—but to me he admitted he was also scared. So they cuddled and kissed for a while longer, then Mason left.

The next afternoon, Jeannie texted him to ask whether he'd like to hook up again. This time he stayed sober, but he was still anxious: What if he couldn't get it up? The more he thought about that, the more anxious he became. When he tried to put on a condom, he lost his erection again.

"You don't have to be nervous," Jeannie said.

Mason hedged: "It's just the aftereffect of the weed from yesterday," he told her. No biggie. Or, he said, maybe it was a "contact high" from being around some friends who'd been dabbing earlier in the evening. He knew the lie was lame. But Jeannie seemed so confident, so comfortable in her body. She could talk about her desires and her limits. She was emotionally open and expressive. It scared the crap out of him. "I thought, *Damn! This girl's got everything figured out!*" he said. "And I'm over here trying to get over the trauma of years of porn usage, trying to differentiate between what I like in the fantasy world versus what I actually enjoy in real life."

A week later, Mason went to see Jeannie act in a play on campus. Again, they went back to her dorm. This time Mason declined to have intercourse out of supposed gallantry: she had another performance the next day and he said he didn't want to keep her up too late.

"Can I ask you a question?" she asked. "Have you ever done it before?"

In the past, Mason would lie in response to that, whether in bed with a partner or, as had recently happened, during a conversation among friends at work. "As a guy, you feel like you should already have had that experience," he told me. "And if you're a virgin, you should dodge the question as much as possible." With Jeannie, it was somehow different. "I decided to just say it," he recalled. "I told her, 'No. I've had sexual experiences but not sex.'"

"Is virginity something you value?" Jeannie asked. "Do you want your first time to be significant?"

Mason thought for a second. "Not really," he said. "I'm happy to have sex if the situation presents itself, but it's not my goal when I meet a girl. It's more like, 'Is she a cool person? Do I enjoy hanging out with her?'"

They continued talking for over an hour; Mason confided in Jeannie about his anxiety, his difficulty expressing emotion. When they began kissing again, suddenly everything clicked. "It felt like this storybook moment," Mason recalled, "because of having an hour conversation beforehand, being able to let my guard down. It's like, after having that conversation, I felt extremely comfortable being with her. Whatever nerves had affected me the previous times disappeared. And I realized: if I can't be fully vulnerable, mentally and emotionally, it stops me from being able to be vulnerable physically. Because, the naked body, that's a very vulnerable thing, you know?"

Mason looked at me earnestly, and I nodded. It certainly can be. And that's pretty different from what he'd learned from porn or mainstream media. "It sounds like you really connected," I said.

"Yeah," Mason replied. He doesn't know what's going to happen with Jeannie; he doesn't know what he wants to happen. The relationship, such as it is, is young. Maybe it will blossom into something more enduring, maybe it will develop into a consistent hookup, or maybe it will fade away, undefined. Regardless, he's grateful to have met her. Because now, he says, he knows what he's looking for—and he knows he won't find it on a video screen.

CHAPTER 3

Are You Experienced?

Life and Love in a Hookup Culture

The tiny dorm room was packed with freshmen, the temperature rising along with their voices and the chemical tang of someone's musk body spray. A raucous group played rage cage—a fast-paced drinking game involving Ping-Pong balls and a whole lot of beer—on a table that bisected the room. Students poured vodka shots into *Despicable Me* Dixie cups that had been set out on a desk. The bladder from inside a box of wine was making the rounds, too: the idea was to slap it (for reasons no one could explain), then chug from the spigot as everyone counted: *one, two, three, four, five, six, seven, seven, SEVEN!* "Oldies" from the mid-2000s blasted from borrowed speakers: *Pussy on my mind, tighter than a headband.* Although this was a midsize college on the West Coast, we could have been anywhere: it was a typical Saturday night pregame, the party before the party, where students get pumped up and a little wasted in preparation for heading over to frat row. I had been invited by one of the hosts, Iris, a fan of *Girls & Sex*, to observe the event. Most of the students ignored me, though a few, assuming I was a professor there to do research, struck up conversations.

A girl in a striped crop top, leggings, and sneakers barreled over, already a little tipsy. "It's so unfair to have a vagina," she complained, once I'd explained my presence. "They're so hard to deal with. For guys, it's just simple. I mean, I've read Judith Butler. I love Judith Butler. I know gender doesn't exist, but still. Girls have to deal with sexism and then on top of it we're so hard to please sexually."

I asked how she thought the boy standing next to us might respond to that. She tapped him on the shoulder. "Do you think as a guy you see the world differently?" she demanded.

He stared at her, confused. "Um, I think as an individual I see the world differently?" he said, then turned his back on her.

"Well," she said, disgusted, "that was a cop-out." She wandered off; when I spotted her again, a few minutes later, she was leaning against a wall, making out with a guy whom, I'd later find out, she had just met.

I watched a different boy, wearing a T-shirt emblazoned with the word "apathy," weave through the crowd. He caressed several girls' bare shoulders, dropped his head on another one's chest (she immediately took a step back), hit up someone for her Snapchat. Iris had invited this boy specially; they'd hooked up a couple of times recently and she thought there was potential for something more. Judging by his behavior, though, that wasn't looking likely. He paused next to me, Solo cup in hand, surveying the room. Iris had told him I was writing about hookup culture; he took that to mean I was focused on rape. "Guys are a lot worse at those Big Ten schools," he said. "That's where you see all the assault. It's awful. Here, guys are *respectful*."

"What about Brock Turner?" I countered. "He went to Stanford. That's not a Big Ten school."

"Yeah," a second boy broke in, "but he was an *athlete.* He wasn't there on merit."

A third boy agreed. "It really comes down to communication," he said. "You have to *communicate* with your partner."

I turned to a girl who had been listening without comment. "Do boys usually 'communicate' with you during a hookup?" I asked.

She grimaced and shook her head.

"I mean, you're usually drunk, right?" I asked the boy.

"Well," he said, "I don't hook up with anyone drunk that I wouldn't get with sober."

"Does 'communicating' also mean you don't just stop talking to a girl after a hookup or ignore her when you see her again?"

He shifted uncomfortably. "Well, um . . ."

"Or that you make sure your partner has an orgasm whether or not you do?"

He glanced longingly toward the group playing rage cage. "Um . . ." he repeated.

The girl interrupted: "The last guy I hooked up with didn't know what a period was," she said. "I told him I thought we wouldn't be able to have sex because I was on my period but then it turned out there was no blood so it was okay. And he was like, 'Does there have to be blood?' " She rolled her eyes. "What does he think it is, a state of mind?" She burst into laughter, taken by her own cleverness.

"But you still had sex with him," I said. "Did you enjoy it?"

She seemed startled, as if I had broken an unspoken rule. "No," she said, no longer laughing. "No, I didn't. Not particularly."

Anatomy of a Hookup

There are two contradictory trends identified in reports about young people's sex lives: one is that they are virtually celibate, too busy playing *Fortnite*, watching porn, scrolling through Instagram, or otherwise living screen-mediated lives to actually connect with another human being. The other is that "hookup culture" along with a plethora of Tinder-type swipe apps have made sex so accessible that everyone is bed-hopping in a nonstop, booze-fueled bacchanal. The truth lies somewhere in between. High school and college students are, in fact, having less vaginal intercourse than they were twenty-five years ago (the studies quoted in the press, though, don't ask about oral or anal sex, both of which have become more common), but that's partly because the context in which they indulge has shifted. In a relationship, couples tend to have intercourse regularly; students who engage primarily in hookups, even those they consider "consistent," do so only sporadically—an irony, given the dissolute presumptions about hookup culture.

A moment to define terms: "Hookup," a word high school and college students bandy about incessantly, is intentionally vague. Although it presumes a lack of connection and commitment, beyond that, the details are indeterminate, shifting with age, geography, and personal experience. I once heard college freshmen debate whether a dance floor make-out (DFMO) "counted" as a hookup. It would have in high school, they concluded, but not anymore, though it definitely would if you engaged in the exact same behavior back in someone's room. In reality, around 35 to 40 percent of college hookups include intercourse, which means 60 percent or more do not; about 13 percent involve oral sex (mostly performed by girls on boys); 12 percent consist of some naked genital touching; and 35 percent—over a third—go no further than kissing and groping. Because of that ambiguity, however, students tend to radically overestimate what their classmates are up to (not to mention allowing others to draw inflated conclusions about their own exploits); that can fuel feelings of inadequacy and FOMO, contributing to pressure to keep pace through undesired sex, coerciveness, or aggression. According to the Online College Social Life Survey, which encompassed more than twenty thousand students nationally, close to three-quarters of both male and female students will hook up at least once by their senior year, engaging in some combination of the above behaviors. The average number of partners? Seven to eight. That's maybe one a semester—not exactly the fall of Rome. Forty percent hook up fewer than three times during their college career, and a full quarter *never* hook up, though 20 percent do hook up ten times or more. The behavior is most common among affluent, white heterosexuals, and predominant in Greek life.

Uncommitted sex in college is certainly nothing new. The real shift, then, is not "hookups" but hookup *culture*: the idea that casual sex is no longer an exception, but that physical intimacy is expected to be the precursor to emotional intimacy rather than its product. Students, both in high school and college, see hooking up as the first step toward a relationship, although most hookups don't result in one. No wonder as many as 85 percent of college students report ambivalence or unhappiness with hookup culture and one in three say their intimate relationships have been "traumatic." According to Lisa Wade, an associate professor of sociology at Occidental College, students express "a deep, indefinable disappointment" in their sexual experiences. "They worry that they're feeling too much or too little. They are frustrated and feel regret, but they're not sure why." Although some students thrive, most wonder why they aren't having fun yet.

"The sex can feel like two people having two very distinct experiences," observed Andrew, a second-semester freshman in Los Angeles who had hooked up with ten girls since school began and had intercourse with five. "There's not much eye contact. Sometimes you don't even say anything. And it's weird to be so open with a stranger." He paused, searching for a way to encapsulate the experience. "It's like you're *acting* vulnerable, but not actually *being* vulnerable with someone you don't know and don't care very much about. It's not a *problem* for me. It's just—odd. Odd, and not even really fun."

To gird against disappointment—as well as a near-fanatic generational fear of the "awkward"—it's crucial to get hammered. Hence, the pregame. To say that hookup culture is lubricated by alcohol would be a gross understatement: it is *dependent* on binge-drinking to create what Wade calls the "compulsory carelessness" necessary for a hookup. Alcohol is, above all, what establishes a couple's indifference: hooking up sober is almost by definition *serious.* Inebriation itself—"I was so drunk"—can even become the reason (or the excuse) for an encounter, as opposed to, say, attraction, interest, or connection. "In the stairwells of my dorm," Andrew told me, "people will talk about how if you didn't black out, you didn't go out."

Most of the guys I met were acutely aware that sex with an incapacitated person is assault. Yet, since you need to be drunk in order to hook up, the trick becomes being (and finding someone who is) wasted enough to want to do it but sober enough to be able to express a credible "yes." And who is to be the judge of that? "I'm very careful about not doing anything if I'm super drunk," a college freshman in North Carolina told me. "I don't want to make decisions I'd regret. So I usually limit it to six or seven drinks. But sometimes more." Even so, he woke up one morning about a month into his first semester with no memory of the night before and a strange girl's number entered into his phone. "I was freaking out, but I texted her and she said she'd had fun, so luckily it was all right."

College anti-assault activists are fond of saying, "Don't tell girls not to drink, tell rapists not to rape." Personally, I don't see it as a zero-sum game. As the parent of a daughter, I firmly advocate talking to young women about the unique way female bodies metabolize liquor: drink for drink a girl will become incapacitated more quickly than a guy who is the same size and weight. I also endorse discussing how alcohol reduces power and obscures judgment, making it more difficult to recognize and escape dangerous situations. At the same time, it's clear that we need to be far more active in discussing how guys' alcohol consumption adversely affects *their* judgment, putting them at risk of engaging in the kind of sexual misconduct that could get them suspended or expelled from school—not to mention harming another human being. Alcohol has been shown to diminish boys' ability to read social cues or notice a partner's hesitation. It gives them the nerve they might not otherwise have to use coercion or force to get what they want; drunk guys are more aggressive when they assault and less aware of their victims' distress. Inebriation also makes boys less likely to step in as bystanders than they would be if they were sober.

Guys in my interviews were less likely than girls to express anger, betrayal, resentment, or feelings of being "used" in hookups. That's partly because hookup culture aligns with the values of conventional masculinity: conquest over connection, sex as status-seeking, partners as disposable. While I've met both male and female students who embrace (or reject) that philosophy, it tended to ultimately advantage boys, an extension of the myriad small, unchallenged ways—in the locker room, in pop culture, in social media, in porn, from their friends, from their own fathers—they learn to see sex as impersonal and female bodies as vehicles for their own gratification. At one end of the spectrum, that sense of entitlement can justify assault, whether premeditated or spontaneous, acknowledged or not. At the other, it typically means indifference toward a female partner's pleasure in a hookup. The Online College Social Life Survey found that between 29 and 53 percent of girls climaxed in their most recent hookup (depending on the combination of acts involved), as opposed to between 56 and 81 percent of boys. That orgasm gap shrinks markedly in relationships, a difference attributed to a combination of familiarity, better communication, sobriety, and—significantly—emotional investment on the male partner's part. In the words of one guy, "It sounds bad, but in a one-time thing, I don't really care." Lisa Wade found that, whether consciously or not, boys signaled a partner's lack of value to them by denying her orgasm and the activities that would most likely produce it (though, at the same time, guys also overestimate women's orgasms in hookups by a third to a half, either out of ego, ignorance, or because the girl faked it). I recalled talking to a high school senior who broke down in tears while telling me that his girlfriend almost never went down on him. "She thinks it's dirty and gross and talks about how it tastes bad," he said. My heart went out to him, it did, especially when he said her rejection made him feel uncared for, unloved. But there was also part of me that recognized the right men feel to sexual pleasure, how dejected and even potentially angry they become when denied it. *After all*, I thought to myself, *you know what an eighteen-year-old girl would call it if a guy wouldn't go down on her? Normal.* There's even some suggestion that, although hookups are explicitly meant to be devoid of feeling, guys in college use them in part to experience emotional closeness, in however attenuated or fleeting a fashion. If that's true, the differential between the sexes in those encounters is even bigger than previously thought: guys derive both physical *and* emotional satisfaction from hookups, while girls generally experience neither.

For some college men, treating a sexual partner—especially one who was not suitably hot or selective—with roughness or disinterest and then bragging about it the next day became a form of image management, a preemptive strike against potential ridicule, the loss of social currency. So, when boys assured me that their friends, their frat brothers, their classmates would *never* assault a girl (it was always those *other* boys), that could feel like a very low bar: having sex that is technically "legal" is hardly the same as sex that is ethical, mutual, reciprocal, or kind. "Casual sex can be great," a sophomore at a Los Angeles college told me. "But you can forget to treat the other person as a human being."

Boys felt less strongly than girls that the game was rigged against them; still, few felt the competitive, detached nature of hookup culture fully served them. They struggled not only with unexpected feelings of connection and vulnerability, but with other emotions: inadequacy, anxiety, insecurity, confusion, disappointment, embarrassment—none of which, as guys, they felt permission to express. That was especially true in high school, where hookup culture, while less studied (academics tend to focus on college students), truly begins.

Ws and Ls

Nate, a high school junior from the Bay Area, recalled sinking deeper into the couch, trying to look chill, like he was such an insider that he didn't need to prove it by standing around with the other kids gathered in Nicole's living room. The party was slowly filling up with students from his high school: first the sophomores, then, later, other juniors and a few seniors. The table next to him was crowded with handles of liquor, beer bottles, and a fruit-flavored malt drink called Great America that was packaged like moonshine, in mason jars. Kids were drinking shots and moking. Some were Juuling. "People at my school like to get fucked up, because they have so much pressure on them," he told me. "So at parties they just want to fucking destroy their bodies and feel like they have no inhibitions. There is a lot of getting drunk and there's a lot of 'fun' I don't completely understand." Nate described himself as in the middle of his school's social scene: friends with the "popular" kids, but also with the "lower" kids. "But the hierarchy really centers around who has access to the parties," he explained. "And I do get invited to all the parties." Nate didn't drink much himself, and he never got high: he wasn't morally opposed to it; he just didn't like the feeling of being out of control. So he usually ended up as designated driver, or taking care of the guy who was puking on the sidewalk, or "protecting" a female friend from some wasted, creepy guy's advances, slinging an arm around her shoulder as if making a claim.

He wandered into the kitchen, where his friend Kyle, who was *definitely* cross-faded (that is, both drunk and stoned), stood on a chair aiming the contents of a can of Sprite at a shot glass on the floor below and, much to the amusement of everyone around him, missing entirely. Nate took out his phone and made a Snapchat video.

At sixteen, reputation meant everything to Nate, specifically his reputation as a guy. Certain things, he explained, could cement your status: playing sports, posting funny videos, getting drunk, and, of course, hooking up. "The whole goal of going to a party is to hook up with girls and then tell your guys about it," he said. "And there's this race for 'experience,' because if you get behind, then by the time you do have the opportunity to hook up with a girl she'll have hit it with, like, five guys already. Then she's going to know how to do things and you won't and that's a problem if she tells people you've got floppy lips or don't know how to get her bra off.

"So, yeah, if you have a girlfriend, that's okay," he continued. "Maybe that's a nine out of ten. But if you can hook up with random girls? That's a ten out of ten. So it's like you *need* to be hooking up to be a man, and you need to be *good* at hooking up to be a man. But how am I supposed to learn how to be 'good' at hooking up without hooking up?"

Nate was a lanky boy with dark, liquid eyes and hair that resisted all attempts at taming: not the best-looking guy among his classmates, maybe, but certainly not the worst. Still, he'd hooked up with only three girls since ninth grade—making out for maybe ten or fifteen minutes with each of them, maybe lifting their shirts—and none had wanted a repeat. That had left him shaken, worried about his skills. "I'm afraid of intimacy," he told me earnestly. "It's a real self-esteem thing."

"Intimacy" was, perhaps, a poor choice of word; it would be more accurate to say that Nate was uncomfortable having drunken sexual interactions with girls he did not know or trust. Yet that was the only form of "intimacy" that genuinely counted among his friends. While girls struggled to find the magic middle ground between "prude" and "slut," boys were pushed to be as sexually active as possible, to knock out their firsts regardless of the circumstances or how they felt about their partners. It was all about checking the right boxes, credentialing oneself. "Guys need to prove themselves to their guys," Nate said. "So to do that, you're going to be dominating. You're going to maybe push. Because it's like the girl is just there as a means for him to get off and a means for him to brag. That's what you've been taught by your friends."

Before the start of this school year, Nate's "dry spell" seemed to be ending. He was in a relationship with a girl that lasted a full two weeks, until other guys told him she was "quote-unquote 'slutty' "—their word, he assured me, not his. Because although any hookup might be marginally better than no hookup, you only truly got points for hooking up with the right kind of girl. "There's this whole thing that if you hook up with the girl that's below your status it's an L," Nate explained. "A loss. Like a bad move." So he stopped talking to the girl, which was unfortunate: he'd really liked her.

Bored with watching Kyle trash the kitchen, Nate retreated to the couch, letting people swirl around him, occasionally taking what he hoped were artsy photos of the trees and lights outside the living room window. He was starting to relax, to enjoy himself. Gucci Mane, newly out of prison, was streaming on Spotify, boasting about sex and money. *Lil mama wanna suck me and she ain't never met me. . . .* The kids at his school, who were mostly white and affluent, ate that shit up.

Suddenly, Nicole, the party's host and a senior, plopped onto Nate's lap, handing him a shot of vodka. Nate was surprised: usually, if a girl wanted to hook up with you, she'd go through an intermediary, instructing her friends to ask you whether you were up for it. There were texts and Snapchats, and if you said yes, it was on—everyone would be anticipating it, and everyone would expect a postmortem the next day. Nate was impressed, if a little confused, by Nicole's boldness. She was blond and slim, with amazing eyes and major breasts. Kyle, who was still splashing Sprite all over the kitchen floor, had been *obsessed* with Nicole. Hooking up with her would be a W—a win for Nate. A big one. And sure, he also thought Nicole was hot, though he never actually *liked* her very much and he'd never been especially interested in her before this moment.

He glanced around the room subtly, wanting to make sure, without appearing to care, that everyone who mattered, everyone "relevant," saw what was going down. A couple guys gave him little nods. One winked. Kyle came in from the kitchen and slapped him on the shoulder. Nate feigned nonchalance, like nothing out of the ordinary was happening. Meanwhile, he told me later, "I was just trying not to pop a boner, because that would be weird. But, you know, I'm pretty sure she felt it."

"Hey, Nicole," Kyle said. "Do you want to show Nate your room?"

"Okay," Nicole agreed. Taking Nate's hand, she led him along a hallway and down a flight of stairs, laughing, smiling at her friends. He stopped when they reached her door. "Are you sure you're sober enough do this?" he asked.

"Mm-hmm," Nicole said, pulling him forward and flipping on the light. Her bed was shoved into a corner next to a window. Outside, a group of kids were smoking weed on a deck. No question: they could definitely see in. Nate panicked. He was already nervous, already afraid that he didn't know what he was doing, already aware that everyone upstairs was going to want to hear the details, but now it seemed like people might actually *watch* them. This was all starting to seem like a terrible mistake, but there was no way to back out. He got through the inevitable, awkward moments where you actually have to *talk* to your partner, then, finally, they started kissing. In his anxiety, Nate bit Nicole's lip. Hard. "I totally messed up," he said. "And I was thinking, *Oh God! What do I do now?*" But he kept going. He took off her shirt and undid her bra. He took off his own shirt. Then she took off her pants. "And that," he said, "was the first time I ever saw a vagina. I did not know what to do with it." He recalled that his friends had said girls go crazy if you stick your fingers up there and make the "come here" motion, so he tried it, but Nicole just lay there, unmoving. "Guys will call girls like that a 'fish,'" Nate said. "Like a dead fish. So I guess if I talked in those terms, maybe that's what I'd say happened." He didn't ask what might feel better to her, though, because that would be admitting ignorance, and he wasn't about to do that.

After a few more agonizing minutes, Nicole said, "I have to go see what's going on upstairs," got dressed, and left, Nate trailing behind. They rejoined the party and immediately went their separate ways. One of Nate's friends high-fived him and slapped him on the back. Another handed him a bottle of Jack Daniel's and Nate took a long slug. A boy who had hooked up with Nicole at a different party fist-bumped him. "Yo, bro!" he said, "We're Eskimo brothers!"

"And the whole time," Nate said, "I'm thinking, *Holy Fuck! I don't think that went very well. But I guess it was experience. So . . . good for me!*"

Then he heard a senior, a guy Nate considered kind of a friend, loudly ask Nicole, "Why would you hook up with *Nate*?"

She giggled. "Oh, I was drunk!" she said. "I was *so* drunk!"

They were calling him an "L."

"I'm sorry, man," one of Nate's friends said, shaking his head.

Nate brushed it off. "No, it's okay, don't worry about it." He went back to the couch and pretended to be drunk to avoid further conversation.

That night, Nate slept over at Kyle's. By then he was completely sober, though Kyle was still stoned. He peppered Nate with questions about Nicole's body and what, precisely, the two of them had done. "I tried not to answer too specifically," Nate said. "I'd just say, 'Yeah, man, it was dope. I feel great. It was totally awesome.'"

They looked through Nicole's Instagram feed and talked about how "great her tits were." "Dude! You hit that!" Kyle said. "Great job!" Kyle's enthusiasm lifted Nate's spirits: he hadn't enjoyed the hookup, but at least it hadn't been a disaster.

Or so he thought. By Monday morning, Nicole had spread the word around school that Nate was "bad" at hooking up: that he'd bit her lip, that he didn't know how to "finger" a girl. That his nails were *ragged*. Maybe she was trying to preserve her own reputation—to avoid being teased for hooking up with an "L." Or maybe she just thought the story was a juicy one. "The stereotype is that guys go into gory detail," Nate said, "but a lot of times it's the other way around. Guys will brag, but they're not so specific. Maybe they'd say, 'I fingered her,' or maybe they wouldn't even say what they did, just that 'we hooked up' and leave it at that. But girls? They'll go into every detail with their friends. What his penis looked like. Every single thing that they did. How it felt. It's like the opposite of what you think."

Nate said he felt "completely emasculated," so shamed that he told his mom he was sick and stayed home from school the next day. "I was basically *crying*," he said. "I was like, '*Shit!* I fucked up!'"

No question, gossip about a guy's poor "performance" could

destroy a reputation. So the guys I talked to actually *were* concerned with female satisfaction in a hookup; they just didn't typically define it through orgasm. Rather, they believed it to be a function of their own endurance and, to a lesser extent, penis size. Andrew, the college freshman in Los Angeles, recalled a guy in high school who "had sex with some girl for the first time and she told everyone he'd ejaculated really quickly. After that, he got the nickname Second Sam. That basically scared the crap out of all the other guys." The public nature of what should be a private act, the invisible audience that was always in the room (apparently holding a stopwatch), could turn boys' early experiences from explorations of sensuality into referendums on masculinity. Guys who'd only had penetrative sex once would anxiously ask me whether they were "premature ejaculators" because they hadn't had the purported stamina of porn actors (whom, they did not realize, maintain their reputation through editing, ED drugs, and penile injections). A college sophomore from Boston explained that he'd developed the habit of glancing at the clock when he started intercourse. "I'd think, *I have to last five minutes, that's a minimum*," he said. "And once I could do that, I'd think, *I need to get to double digits.* I don't even know if it's necessarily about your partner's enjoyment. It's more about getting beyond the point where you would be embarrassed, maintaining your own pride. It's about reaching a place where you feel like she's not going to go tell her friends how disappointing it was. But it also turns sex into a task. One that I enjoy to a certain degree, but one where you're monitoring your performance rather than trying to live in the moment."

Eventually, Nate decided he had to take a stand, if only to make returning to school bearable. "So I texted her a long message and said, 'I'm sorry that you didn't enjoy it, but I would never do that to you.' Then she felt really bad. She stopped telling people, but we were really awkward around each other after that. It took me until the next semester to recover, until I got a real girlfriend."

The Morning After the Night Before

"You should write about texting after hookups," Ellie said. "There's norms, but no one really knows what they are."

It was ten o'clock on the morning after the night before, and a group of freshmen in various stages of postparty fogginess had regrouped in Iris's dorm room. Iris, dressed in a business suit and bunny slippers, was rushing around, getting ready to leave for a debate tournament. As she dabbed on makeup, she debriefed her friends about a guy she'd made out with at the previous night's party, whom she sort of liked. "He said I should come by his frat some time," she said. "So—maybe . . ."

Ellie, the girl who'd gone on about Judith Butler, watched groggily from bed, still in her pj's, along with another friend, LeeAnn, who clutched a thermal coffee mug. Caleb, who lived down the hall, seemed the most chipper of the bunch. He hadn't gone out the previous night because, he said, he'd been drinking too much lately and thought he should cut back. A few weeks earlier, he'd visited a friend who was in a fraternity at a nearby state university; one night of that was enough liquor for a long time. "I saw what bro culture really is," he said. "I mean, individually, they weren't douche-y people; they're all really friendly and nice and smart. But after a couple of hours of playing video games you realize that five guys have literally gone through five handles of vodka and it becomes like night and day. I actually don't remember much, but I know we were walking to the party, having a great time together, then the second we got in the door they all dispersed to find a chick to talk to, and that was it. I didn't see any of them again for two hours. It was so clear-cut: you get there, you're smashed, you try to hook up, and then you try to remember it in the morning." Perhaps, as Caleb did, you also wake up encrusted with your own vomit.

The sex in those hookups can't be too good, I commented.

They all nodded in agreement.

"Sometimes," LeeAnn mused, "I've gone through with hookup sex because I feel—'obligated' is a strong word, but it's kind of *expected.* I feel like the other person would be bummed out if I didn't."

"There would be something lacking," Ellie agreed.

"Just the *expectation* that they were going to get laid," LeeAnn added.

I turned to Caleb and asked whether he was aware that girls sometimes hook up out of duty rather than desire.

"That is really surprising," he said. "I did not know that. I thought if you're going out and getting drunk and trying to hook up with somebody, you put yourself in that situation so you can get your own pleasure out of it.

"I do feel like guys are oblivious to what goes on for girls in hookups, though," he added. "As a guy, the way you think about a hookup is very binary—either you overthink it a ton because you care, or you dissociate yourself from it entirely and do it to, like, get praise from your friends or something. I want to believe I'm self-reflective, but a lot of time, yeah, it's either overthinking or not thinking about it at all."

Which brought us back to the subject of that all-important post-hookup text. According to sociologist Lisa Wade's research, partners are expected to be *less* friendly after a hookup than they were before, at least for a while, as the last step toward affirming that requisite meaninglessness. Girls used to tell me that the first person who admitted to "catching feelings"—a phrase that makes affection sound like an STD—"loses." But that creates a predicament for the over half of both men and women who say they'd like their hookups to turn into something more. "You meet someone for the first time, hang out with them the whole night, and potentially end up hooking up with them," said one of a group of fraternity brothers I talked to at a Midwestern college. "Then the next day you're walking down the street and it's like . . ." He looked up as if spotting someone he knew, then quickly averted his gaze. "There's *no* eye contact. It's very much like, 'I'm not saying hi. We're just going to move past each other.' Like, 'We're not actually *friends.* It was just this thing that happened at a party.' I hate it."

I asked him, in that case, what kept him from breaking protocol. "There's this sense of—I don't know, maybe weakness? You're trying to play it cool. Like"—he switched to the laconic "bro" voice I'd heard so often—" 'Oh yeah, it's cool. Whatever.' I mean, I don't know if she was into it or if she's trying to move on; she doesn't know if I'm into it or if I'm trying to move on. And because of that question mark, you don't want to make yourself vulnerable and get shut down for it and then be like, 'Oh, you idiot. It was a party! You should've realized it was a one-time thing!' So you just play it safe and don't say anything."

"Even if that means you miss the opportunity for something more?" I asked.

"Yeah," he said. "Even if it means you miss the opportunity for something more."

Back in California, Caleb echoed those sentiments. "I hate

texting girls. It's more anxiety-provoking than the actual hookup. There's this whole game you play with the timing. How long do you wait to send the first text? How long do you wait to reply?" It's all fraught with layers of meaning, with projecting the right balance between casual and clingy. "And after you get over the timing, you have to worry about the actual content," he continued. "What you say, yeah, but also using 'you' versus 'u' or purposely misspelling a word, like 'wanna' instead of 'want to.' And then, you're Snapping back and forth with someone and they suddenly take twenty minutes to respond—"

"Then you have to take twenty-five," Ellie interrupted.

"Exactly," Caleb said. "And it's like, 'I *know* you have your phone right there! We all always have our phones on us. It's not like you didn't see the Snapchat.' So then you're left staring at it waiting for them to reply."

"Plus, you don't know what anything means," Ellie added. "There's no inflection. Like, if someone texts you, 'Thanks for a great night.' That could mean: 'I want it to happen again,' or it could be closing it off. You just don't know."

"What about when a guy slides into your DMs on Instagram?" said LeeAnn.

"That's totally different than a text," Caleb said. "A DM is more like you *know* why you're talking. You're going to hook up. DMs are the most casual. Anyone can DM you. If you're really interested in a girl, you'll text. *Maybe* you'll Snap."

Nearly every interview I've ever conducted on hookup culture, like this one, quickly devolved into a discussion of its drawbacks. Why, I asked the group, do you continue to participate?

"Well, it is a lot easier," Ellie said. "And it can be fun."

"And it's still sex," Caleb added. "Even bad sex is still sex."

The Feminist Fuckboy

"I do believe, though, that you can have a casual sexual lifestyle and also be respectful and loving." Wyatt was entering his junior year at a small liberal arts college on the East Coast that, despite having no Greek life, was known for its entrenched hookup scene. In part, that was due to a dearth of men on what was formerly an all-women's campus: competition for male attention was fierce. Added to that, Wyatt was a rare heterosexual male dance major and handsome—wavy chestnut hair, sculpted features, a well-muscled body that he was only too happy to flaunt at parties, removing his shirt at the least provocation. He pretty much had his pick of partners on campus, and over the past year, he had taken full advantage of that. "At my school, they call it 'Golden Dick Syndrome,'" he said. "There is this inherent godliness that straight men tend to feel here, where it's just like, 'Everybody wants me.' It's just the way it is. Instead of one girl across the party that you know is interested in you, imagine there are *six*. It can actually feel kind of weird."

Wyatt had not been popular when he was younger. As he said, "Most people who go to colleges like mine weren't exactly the kings of their high schools, you know?" He was more the guy who stayed home on Saturday nights playing *Grand Theft Auto* and listening to hip-hop. "I'm a big Kanye defender," he said. "For how hypermasculine he is, he actually strips it down more than most big stars. He talks about what a jerk he is and how he's a sex addict and has depression and anxiety. In the end, I think he's a lot more up-front than most men."

For much of his freshman year, Wyatt had a girlfriend—his parents had just separated, which had been a huge shock, and he clung to the security and emotional support of a relationship—but since that ended he described himself as being "on the prowl." Every weekend he would have sex with a different girl (in one-offs, he said, he "nearly always" wore condoms). As soon as it was over, he'd be back on Tinder or stalking someone on Instagram, thinking about who to pursue next. "The number of women I've slept with is unbelievable," he said. "I can't even begin to count them. I don't dare." Sex made him feel good: good about himself, good in general. It eased anxiety, numbed depression, distracted him from and substituted for any true feeling. And, of course, it was fun. "I like that hookups are not this huge emotional thing," he said. "We can just be primal." But he was starting to worry that he might be investing too much of his self-worth in sexual conquest. "My actual attraction to the person almost doesn't matter," he told me. "If she is assertive enough, there's a very good chance I'll just say yes."

I'd met Wyatt through one of his former high school teachers, who'd told me that he led workshops on consent for men on his campus. That was true, he said, although he found other guys were not always so keen on the message. "They'll say, 'It's just not how it's supposed to be.' Or they offer up all these very specific circumstances like, 'Let's say this girl starts grinding on me and kissing me. Do I still have to ask her?' They pick away at it. So I find myself explaining, 'Why would you take the risk?' Just deal with the fact that it's awkward and maybe you don't think it's sexy. I'd rather have something not be sexy than be accused of rape."

Personally, Wyatt saw establishing consent as erotic, part of the buildup and play of sex. His MO was pretty straightforward: after hanging out and talking to a girl for a while—maybe drinking at a party, though sometimes sober—he'd make his "big move." "I'll say, 'Hey, I think you're really pretty. Can I kiss you?'

It's direct, and I don't want to do that thing where I lean in and get rejected like in the movies. And as things progress, I ask, 'Do you think that you're sober enough? Do you want some water?' Stuff like that. Things that I would want someone to do with me."

He was scrupulous about checking in at every escalating stage of an encounter, too, but similar to the boys he counseled, his female partners haven't always reacted well. "Some of them don't like it," he admitted, "but they don't stick around long. There was one girl this year who I was very attracted to. Very excited to be hooking up with her. I asked, 'Can I kiss you?' and she started laughing. She thought it was 'cute,' but she was also like, '*What?*' Then, I asked, 'Can I take your shirt off?' And she was like, 'You don't have to ask!' That just makes me mad because it is exactly the quagmire I want to avoid."

For every girl who's seemed put off, though, more have been appreciative. "I'll just put it out there," he said, "affirmative consent is really hot. It's exciting to have a girl saying, '*Yes!* I want you to do this.' '*Yes!* I want you to do that.' To feel she's really into it. It's a pretty awesome thing for both of you to have that sort of connection."

Wyatt said he's also clear with his partners that he's not looking for anything *more* than sex, even if he wants to see them again. "I'll say, 'It's going to be in a very casual capacity, and if you can't do that, then we can't do this.' " When, on occasion, he's sensed someone starting to "catch feelings"—wanting to know more about him, asking questions, maybe wanting to do something together outside of the bedroom—he's "floated away," growing "flaky" and "distant." "Because I'm not there for that," he explained, "and I'm not going to be there for that."

From my interviews with girls, I knew that some of them would be fine with such an arrangement: they, too, were only looking for a wild night and a warm body. But others felt that hookup culture had trapped them into a game of emotional chicken: they had to feign a disinterest they didn't necessarily feel in order to avoid appearing "clingy" or desperate for a boyfriend.

So I asked Wyatt: Do you think the girls that you were with when you got "flaky" are the same ones who would tell me that all guys are "dicks?"

He laughed uncomfortably. "Gosh! I mean, I hope not, but . . . yeah." He winced. "Yeah, that could be true. I mean . . ." He let out a sigh. "Yeah. That could be true. Who knows what girls really think when you tell them you don't want to be with them, you know? Maybe they do tell their friends that all guys are the same. That if you try to move forward emotionally, guys immediately back off . . . because"—he winced again—"I've done all that. I have. But in my defense, I always try to ask, 'Are you sure you're okay with the fact that I don't go to lunch with you?' "

What would happen, though, I asked, if when a guy said that, the girl responded that she *did* want to go to lunch? What if she said that the truth was, she was looking for a loving relationship?

"Well, at my school that would be lethal for her social life."

So, I pressed, what choice does she really have, then? If you're having sex with girls you're not especially attracted to or interested in and your supposed righteous honesty about not wanting a relationship potentially forces them to either deny their own feelings or sacrifice their sex lives, how, exactly, are those hookups "loving and respectful"?

Wyatt nodded emphatically. "That's what I'm trying to say! I haven't always been true to my own philosophy. It's been sort of—masturbatory on my part."

That's why, he continued, he, too, was starting to sour on hookup culture. Over the summer, Wyatt had been hanging out a lot with an old high school friend in the Bay Area, going to bars and parties. At the end of every evening, he would invariably have sex with one of the guy's female friends—a different one every time. Finally, after Wyatt indulged in a drunken threesome with two girls, the other boy confronted him: "I go out with you to have a good time together, but it feels like you've made a checklist of all my friends so you can go through it and fuck them and then discard them." That wasn't far from the truth, and it stung. "I realized I was becoming, like, a feminist fuckboy," Wyatt said. "The kind of guy who says all the right things, but still treats women badly. And that feels horrible. . . ." Wyatt broke off, pausing for a long time. "No," he continued quietly. "That's a lie. In the beginning it felt amazing, but eventually not, because there is no investment, because the sex doesn't mean shit to me. Because the other person doesn't mean shit to me. And, well, I'm not going to lie. I've liked it. I thought I was okay with it. But I'm starting not to be okay with it. Because . . ." He paused again. "Because there's more to me than that."

As a journalist, I don't prescribe the circumstance in which young people ought to have sex: my job is only to describe the context and explore its impact so they can make educated choices, maybe disrupt the conventional script. For girls, I concluded that a hookup was likely to give them a feeling of being wanted or desired for an evening, an adrenaline rush, a war story to share with friends. It was less likely to result in good sex or help them develop the tools they would need for either good sex or emotional intimacy. After hearing from dozens of boys, I would say the same to them, with the additional qualifier that hookup culture presumes that they, unlike girls, lack even a basic capacity for love, that they neither can nor should acknowledge emotional vulnerability—not in others, not in themselves. Other cultures give boys more credit than that. In comparing Dutch and American families' attitudes toward teen sexuality, for instance, sociologist Amy Schalet found that parents in the Netherlands considered boys to be both capable and desirous of emotional connection; US parents, by contrast, dismissed young men as "driven by hormones" and only interested in sex. Perhaps not surprisingly, although boys in both countries overwhelmingly said they wanted to combine love with lust, only the Dutch saw that as normal: American boys each thought his perspective was a personal quirk, unusual among his peers. Yet, a large-scale survey of high school students found our boys were as emotionally invested in their relationships as girls; perhaps having had less practice or support in sustaining intimacy, though, they were less confident in navigating them.

The boys I met felt at least as isolated in their struggles over love as they were about sex and, if anything, were more hungry to discuss it. There was the twenty-year-old at a Big Ten university who had been dating the same girl for three years. His fraternity brothers as well as his father were pushing him to cut her loose so he could take advantage of the school's infamous party scene. "My dad says, 'I'm not paying fifty grand a year for you not to get laid,'" he told me. "Hearing that from him—it doesn't make me second-guess what I'm doing exactly, but it hurts that he's not fully behind me."

Or another guy, a college sophomore in Chicago who told me he hadn't much enjoyed the three hookups he'd engaged in (and had ghosted all three partners afterward) but didn't know how else to show his interest in a girl. "The thing is," he said, "I could never ask a random girl on a date. That would just be weird."

So, I said, it would be more appropriate to get drunk, make

out with someone you don't really know on the dance floor and maybe have intercourse with her than to, say, ask someone you like from one of your classes to go to a movie?

"Yeah," he said sheepishly.

"Doesn't *that* seem 'weird'?" I asked.

"Absolutely," he agreed. "I think that all the time."

Or a college junior in Oregon who told me, "I've had two one-night stands in college, and both of them have left me feeling empty and depressed. I have no idea what I gained from those experiences other than being like, 'Yeah, I had sex with someone.' There were no feelings of discovery or pleasure or intimate connection, which are really the things that I value. I mean, what is this dance we're doing right now if all we take away is a number?"

Then there was the college sophomore in Los Angeles, one of the more sexually active young men I met, who fell silent when I asked about the most intimate act he'd ever engaged in, finally saying, almost reverently, "Holding hands."

WHEN I TALKED to Wyatt again, a year later, he had fallen in love with a girl he'd met at a semester-long dance intensive. "She's sociable, bubbly, outgoing," he enthused. "She's very passionate, just like I am. We both love to talk; there's no room for the shallow stuff. We're very much a match."

"So," I asked, "do you go out to lunch?"

He laughed. They do, he said. And to dinner. And on dates. "We have a very healthy sex life, too," he added, "but the difference is, there's an investment beyond sex. She's the only person I've wanted to cuddle with or take a nap with or spend the night with. We both wanted that. We both approached it that way from the beginning. If we were going to do these things sexually, they were going to be earned through intimacy and it was going to matter."

Wyatt said that by the time he met his girlfriend, he'd become fully disillusioned with the "businesslike" nature of hookups; he was tired of treating his partners like objects, tired of feeling like one himself. "Yeah, I was the boy toy," he said. "But having sex with someone you love, with this person who's happy that it's *me*, Wyatt, her boyfriend, not just some body? That's amazing. That's something else."

I was unexpectedly moved by boys who spoke of love, in part because of the stark difference in their description of their female partners. Rather than speaking of them as a collection of body parts (Wyatt never even mentioned his girlfriend's appearance), they would reference a girl's character, intellect, personality, generosity, humor, strength, talent. They spoke with admiration and respect for girls' accomplishments or insight—they veritably glowed. They tended to be more conscious of their partners' sexual pleasure, but they also expressed more satisfaction with their own experience: they felt more relaxed, more trusting, more free. Again, that's partly due to the learning curve involved in being with a new person, but it's also because young people infuse relationship sex with the care and kindness they avoid in a hookup.

"Looking back," Wyatt said, "I *did* have Golden Dick Syndrome. And yeah, I led the consent workshops for guys, but I've realized that consent is the *bare minimum* that should be expected when you're with someone. People think once they've been granted consent for sex that all bets are off in terms of what you do to someone emotionally, how you treat them. And that's not true."

Nate, too, fell in love shortly after his ill-fated junior year hookup. His girlfriend, who was a year older, was flirting with him in school one day and suggested, "Hey, we should hang out sometime." He was reluctant at first, afraid of being mocked again, but then he told himself, *Nate, you cannot hide from this your whole life*, and asked her out. "She is just amazing," he told me, grinning. "She's so kindhearted. So sweet. We just totally like to be together." And physically? "The first time we hooked up, we had oral sex," he said. "Both ways. She thought it was funny that I didn't know what I was doing, because she did. She'd gone out with an older guy for two years. But she smiled at me the whole time and was genuinely laughing and having fun. I was like, 'You're enjoying this?' It was shocking to me. But she let me know what she wanted, and I let her know what I wanted, and it was all very healthy."

Even a year later, after she'd graduated and gone to college, breaking Nate's heart, he didn't regret the relationship. "I'm learning what love really is," he told me. "And I've realized that I'm so unattracted to the idea of making sex just good for me as a guy so I can brag about it to my friends. That seems so stupid if sex is about being intimate and enjoying a shared experience with someone. And, honestly, I don't think a lot of the guys really even have fun doing it, either."

I am neither a romantic nor a traditionalist, nor am I opposed to uncommitted sex, though I don't consider "uncommitted" license to be unkind. Relationships are not inherently healthy, particularly given how little guidance young people receive about intimacy: romantic partners can be controlling, coercive, even violent—nearly 1.5 million high school students nationwide experience physical abuse from a dating partner each year (and only a third of them tell anyone about it). Breakups can be devastating, especially for teenagers, and no one likes that pain. Still, falling in love, experiencing intimacy, vulnerability, even heartbreak are important steps in adolescent development and crucial practice for successful adult relationships. Hookup culture short-circuits all of that. Nor, in the end, does it deliver on its promise of sex that is liberated, joyous, and free of consequence. Hot sex with a cold heart may seem an expedient choice—all the fun with none of the emotional exposure—but for guys, it can create yet another block to authentic connection. If they don't learn the skills to support and sustain intimacy, they may end up unprepared for mature romantic relationships. Yet, ironically, they're more dependent than girls on those relationships for emotional support. As a result, young men become more distraught when things get rocky, which makes them even more leery of risking their hearts again, perpetuating a cycle of disconnection.

That said, according to sociologist Lisa Wade, one of the lessons both guys and girls often do take from hookup culture is a knowledge of what they don't want, which, while sometimes hard-won, is not without value. As Wyatt said, "I think I had to go through all that. Not to say that I'm glad, but I'm thankful for those experiences. The understanding I have now came from fucking up, from giving in to my impulses.

"Some people can participate in that culture and it's fine," he added, "but it was not great for me. It was not what I wanted to hang my hat on. I want to think there is more to me than that. Sturdier things. Family. A girlfriend I love. Doing something meaningful. These are the things I can hang onto more than how many girls I fucked."

Wyatt and I were catching up via Skype, about to wrap up, when I happened to get a text from Nate. I hadn't heard from him in months; by now he was in the spring of his senior year of high school and was out touring the colleges where he'd been accepted before making a final choice. He was texting from a school in Southern California.

WTF is up with hookup culture? he wrote. *It's like an orgy here—is that the way to live? Should I be investing in that or in forming meaningful connections with women? Maybe I should just go to BONETOWN and then try to be an emotionally available human being afterward? Or can I skip that step?*

Rather than responding directly, I read the text to Wyatt and asked for his advice.

"I would say . . ." He thought for a long moment. "Don't sell yourself out to be a part of that. The hookup scene has a loud voice in college. But I don't believe anymore that it's all college has to be. I have plenty of friends who don't live that kind of life—me included—and they're so much happier for it. So, if you have that gut instinct that it's not for you, man, don't do it. Don't do it just to fit in. That's what I did. And it kills you."

I typed quickly as Wyatt spoke and sent off the result to Nate. Immediately, three dots indicating he was replying popped up on my screen. Here is what he wrote:

THANK YOU. Really. Thank you. Exactly what I needed to hear. This is where my heart is.

And then he added a heart emoji.

I smiled and tapped back a thumbs-up. I hoped that exchange would help him move forward with confidence in who he truly was—an intense, tender, loving young man—rather than denying or devaluing those qualities in order to have some contrived version of "the college experience." I thought about all the other Nates out there. So few teenagers, especially boys, have an adult with whom to talk through their confusion about sex and love, who can help them with basic decision-making.

I turned my attention back to Wyatt, who seemed to have read my mind. "My heart breaks for those young guys who are questioning the hookup culture, who are asking, 'What should I do?'" he said. "It's trial by fire. You've really got to consider how you'll look back on it later. If it's 'That was awesome! I was experimenting!' well, great. But if long-term, it's going to haunt you the way it haunts me now . . . think twice."

CHAPTER 4

Get Used to It

Gay, Trans, and Queer Guys

Zane often referred to the "gentlemen" he'd hooked up with. As in "the wealthy Republican gentleman in the penthouse apartment," whom he met on Grindr, the swipe app for men seeking other men. The word was a nod to his southern upbringing, a soupçon of Blanche DuBois to warm the chilly northern city where he was now a college sophomore. Zane grew up in Tennessee, in a town of eighteen thousand people, nearly all of whom are conservative Christian Republicans. His dad, a factory worker, and his mom, a secretary, taped a portrait of George W. Bush to the family refrigerator when Zane was a child. The other boys in his high school favored camo gear (before it was trendy) and chewed tobacco; they worshipped Jesus on Sunday mornings and football on Monday nights. At his Baptist Sunday school, Zane learned how to "pray away the gay," although at the time, he didn't think that applied to him. Sure, he was more "effeminate" than other boys: his favorite color in kindergarten was pink, and he donned green face paint and a witch's peaked hat on Halloween. Oh, and there was his longtime obsession with singer Stevie Nicks. But he also had crushes on girls and loved—truly loved—his high school sweetheart, even though their relationship lasted a mere two months, and he never did enjoy the physical aspect. They had oral sex once, but it just felt wrong. "You'd think that would be a tip-off," he deadpanned.

So Zane surprised himself when, during welcome week of his freshman year, he "got tipsy" at a party and made out with another boy. "I thought, *Whoa! I kissed a dude!*" he recalled. "And then I went out and kissed a few more!" He began referring to himself as bisexual. Although he didn't want to be one of those guys who used that identity as "a stepping-stone to gay," by spring he'd recognized the truth: he was only sexually attracted to other men.

You might think you know this story—the one about the small-town boy who grew up "different," headed to the big city, burst out of the closet, jumped over the rainbow, and realized that things do, indeed, get better. And in a way, you would be right. But you would also be wrong. Because this is not 1980, or 1990, or even 2000. Attitudes toward sexuality and gender identity have changed radically, if not universally, even in rural southern towns, or at least in some of them.

A few months before meeting Zane, I sat in the audience of a new production of *Angels in America*, Tony Kushner's Pulitzer Prize–winning two-part play about the AIDS epidemic and gay identity in 1980s New York. I'd lived in Manhattan during the era when the play was set, and tears streamed down my face as I remembered friends who had died of "the gay plague." It took enormous courage to come out in those days, to face likely rejection by parents and community, discrimination by employers; to forfeit the legal protections of marriage and any hope of parenthood. How, I wondered, might life have been different for my gay peers (male and female) had they come of age today? Attitudes have shifted more dramatically than could once have been imagined: in 2004, for example, only 31 percent of Americans approved of same-sex marriage; by 2019, 61 percent did. Even among groups traditionally most opposed to the concept, tolerance is on the rise: two-thirds of Catholics and over a third of white evangelical protestants now endorse it. The boys I interviewed could hardly remember a time before wedding a person of the same sex was legal. They grew up binge-watching *Glee* and *Modern Family* and *It's Always Sunny in Philadelphia*. They helped drive the box office for *The Perks of Being a Wallflower* and *Love, Simon* and pushed "Same Love" up the *Billboard* charts. In the corporate world, Apple's CEO, Tim Cook, is openly gay, as, at this writing, are the CEOs of Qantas, Burberry, DowDuPont, and Lloyd's of London; gay men now earn on average 10 percent more than straight men with similar education, experience, and job profiles. In 2019, Pete Buttigieg, the mayor of South Bend, Indiana, and a former naval officer, became this country's first openly gay presidential candidate. As acceptance has risen, the average age of coming out has dropped from twenty-five to sixteen; many children do so even younger and, in general, with more societal support. Twenty percent of millennials—dubbed "the gayest generation in history"—are LGBTQ+; more than half of those identify as a gender other than the one they were assigned at birth. Among Gen Z, only two-thirds say they are "exclusively heterosexual," and a record number reject all binaries (gay/straight, man/woman), preferring the all-inclusive "queer." Some of the gay boys I interviewed pledged frats; some of the trans boys played on men's sports teams. The most conventional, straight, penis-toting dudes spontaneously offered up their "pronouns" when I asked how they identified, although I was usually referring to ethnicity, religion, and sexuality, not gender.

With so much changing it can be tempting to forget what has not, especially for those who would feel uncomfortable talking to a reporter like me. LGBTQ+ teens remain exponentially more vulnerable than their straight peers to mental health crises, drug abuse, bullying, dating violence, and sexual assault, particularly when they have no family support. The 50 percent rise in cases of syphilis and gonorrhea among boys ages fifteen to nineteen since 2011 disproportionately affects those who have sex with other males, and queer boys compose one in five new HIV diagnoses; only 49 percent report regular condom use. African Americans, low-income men, and those who live in the Deep South are at greatest risk not only of contracting HIV but of needlessly dying of AIDS, in part because they fear repercussions for seeking treatment. Abstinence-only "education" (which, in six states, legally prohibits instruction that "promotes a homosexual lifestyle") heightens that stigma, but so does the exclusion of sexual and gender diversity from many supposed comprehensive sex-ed curricula: silence allows unsafe behavior and potential exploitation to flourish; denies young men affirmation and information on how to protect against HIV and other sexually transmitted diseases; marginalizes LGBTQ+ teens among their peers. It also pushes gay boys, perhaps even more than straight boys, into viewing pornography as a how-to manual. Porn is the only realm where sex between two men is consistently represented, practiced, and validated. In that sense, gay porn may be more affirming, more liberating, than that featuring performers of different genders (or than the mainstream "lesbian" porn that is aimed squarely at heterosexual guys). Still, almost by definition, gay porn reinforces the taboo nature of sex between two men. It also traffics in its own stereotypes about race, gender, masculinity, age, and bodies, and over-portrays anal sex (the largest survey ever conducted found that only 37 percent of gay or bisexual American men reported anal intercourse in their last partnered encounter). A quick glance at the most-viewed "gay" clips on Pornhub also reveals a concerning eroticization of "barebacking" (anal sex without a condom), particularly in fantasies featuring power imbalances, such as stepparent/stepson or coach/young athlete. Although far less research is devoted to the impact of porn on young gay guys than on their straight counterparts, it is probably safe to assume that the genre does not encourage greater humanization, mutuality, or caring between partners; given the dearth of more realistic—let alone positive—portrayals of men having sex with men in mainstream media (among the exceptions are the independent films *Moonlight*, *Weekend*, and, improbably, *Wet Hot American Summer*), that, too, is worrisome.

As for Zane? In a way, he resented having to come out to his parents at all. Even if they would be supportive, even if they'd suspected for years, the need to actively disclose his sexuality seemed, to him, inherently stigmatizing. "If I was straight, if I was dating girls, I wouldn't have to do it," he said. "I wouldn't have to sit them down and say, 'Mom and Dad: I'm straight.' I was living openly by then. They knew I had gone to Pride in Nashville. I just kind of wanted them to take the hint."

But that umbrage, he admitted, may have masked a deeper anxiety. Once he told his parents, it would all be so . . . official. "I was still pushing back on being gay in a lot of ways then," Zane said. "It was a weird mental struggle of not wanting it to happen. I didn't just want to become 'the gay kid' from my town. I did well in school; I participated in the community. I didn't want all of that to be erased by being the gay kid who goes to college up north."

Several times during the summer after his freshman year, Zane came home from a late evening out and woke up his mom, asking to talk, thinking he'd tell her—and then chickened out. Meanwhile, he came out to his siblings, his high school friends, other adults. Then, one night, after what he described as "a few too many glasses of wine," his mom began to needle him. "You're different than you used to be," she said. "Something's different, what's going on?"

"I said, 'I don't know what you're talking about,'" Zane recalled, "but she kept on and I kept hearing, 'You're *different*. You're *different*,' so finally I said, 'Why are you telling me I'm *different*? If anything, being at college has made me more myself. I was deluded living here!'"

She kept pushing until he blurted out: "I'm not straight!"

His mom started to cry—not from disappointment, but out of joy that he'd finally confided in her—and hugged her son. "She was like, 'I've been trying to get you to tell me!'" Zane said. "'It's totally fine!' But I was kind of angry: she'd put me through some mild emotional abuse to get there. I said, 'Since I'm telling you things you don't know about me, here's another one: I smoke.'" Then he went outside to calm his nerves with a cigarette while she, in her excitement, went to share the news with Zane's dad, who was asleep. "He gets up for work at five in the morning," Zane said. "So he was like, 'Okay, but why are you waking me up?'" The next day, he told Zane that he couldn't wait for him to bring someone home. "So that," Zane said, "was really beautiful."

News travels fast in small towns, and although one might imagine that deep in rural Dixie, Zane would be shunned, for the most part, the opposite was true. People seemed to go out of their way to be seen talking to him on the street, touching him on the shoulder, saying, with just a little extra heft to their voice, that they were so happy that *he* was happy and living his life. "It was a different tone than when they ran into my friends," Zane said. "The whole conversation was pretty much trying to say, 'I know you're gay and it's okay with me.' I didn't take offense, though it could be tiring." There was, for instance, the clerk at his favorite shoe store who dragged him over to meet another customer, a middle-aged woman whom he recognized as the mother of the only other openly gay boy in town. "It's 'othering,'" Zane said, "but it's also kind of heartwarming. And it's better than the alternative. If they're willing to be this accepting, I can give them room to grow."

So everyone was fine. They were fine with Zane being out. They were fine with him being "somewhat of a twink" in his self-presentation, his speech, his cultural taste. At Christmas, when he stepped off the plane dressed in high-heeled boots and what he called a "Jackie Kennedy–esque peacoat," his nails painted a tasteful shade of greige, his parents didn't flinch (although his father also didn't speak for the entire drive home). That basic level of support, at a minimum, was typical among the gay boys I met. Although some of the youngest ones weren't yet out to family, and a few struggled to stay connected to parents whose religion considered homosexuality a sin, none had been entirely rejected and most felt comfortable somewhere—at least with their peers. For many straight teens, having a gay friend had become a mark of their open-mindedness. A gay friend (although probably no more than one) could even enhance the status of a straight boy, making him appear secure in his own masculinity. Still, that tolerance of a kind of social queerness—the gay sidekick, the theater boy, the bitchy queen, the must-have accessory—is not the same as an acceptance of sexuality. "My straight friends are still uncomfortable with that part," Zane said, "particularly the idea of anal sex."

The gay boys I met had watched their straight friends engage in the ordinary rites of hooking up, dating, falling in love, falling apart. They would play the role of confidant or "Queer Eye," or tag along as a fifth wheel, but they rarely got to be the romantic lead. Part of the problem was that their pool of potentials was so small, especially in high school: just as a pair of five-year-olds won't necessarily be friends simply because they're the same age, the only two queer boys in the twelfth grade may or may not feel the heat. Few had formal opportunities to meet age-appropriate partners. College communities were somewhat larger, more open, but could feel uncomfortably insular. The mainstream party culture, meanwhile, felt anywhere from unappealing to unsafe. "I did a few weeks of the frat scene and I was like, 'I'll be damned if I go back there,'" Zane recalled. "It doesn't feel right for me, going to a party where it's a bunch of straight dudes and a bunch of straight girls. It's very rare to find someone who's gay.

"But that's okay," he continued. "The entire city is at my disposal for meeting gentlemen. And that's where Tinder and Grindr come in."

"A Weight off My Chest"

Devon's new college teammates weren't saying anything different from what guys say in locker rooms everywhere.

"Oh yeah, dude, I'd totally fuck her!"

"It didn't matter what she looked like—the lights were off!"

Still, hearing them, his stomach flipped, his mind careening to high school. "Because back then," he said, "the guys on the team used to say those things about me."

In those days, Devon was still struggling to live as a girl. He'd grown his hair out to shoulder-length from the short crop he'd worn through middle school (when he was considered a "tomboy") and had asked his mom to buy him more body-conscious, feminine clothes, embarrassing as they were to wear. He tried to join in female friends' gossip, agreeing that a boy they liked was cute, or claiming himself to have a crush on someone. He even tried kissing a guy once, though it didn't go well. "But it was a valiant effort," he recalled, laughing.

Devon was also a competitive swimmer, a good one, at a local club. His younger brother was on the boys' team for a while; he hated being in the locker room because of how the other guys talked about his sister. "I had large boobs and I was also skinny, so I was the object of a lot of that misogyny you see in athletic culture," Devon said. "Some of them said things right to my face, talking about my boobs or saying they wanted to eat me out, stuff like that. And I was like, 'I hate all of you.'"

Sometimes, back when I interviewed girls, I'd wish that boys could inhabit their bodies, just for a day, to see what they live with, the perpetual judgment and reduction to appearance. Devon had actually done that—for over eighteen years, and that time as a woman affected his identity as a man. He was twenty-one when we first spoke, a junior in college with dark brown eyes and neatly shorn auburn hair. A shirtless photo he'd posted on Instagram revealed an athletic torso: broad-shouldered with a tapered waist, cut biceps, chiseled abs. He was short for a guy, but undeniably handsome. The only obvious signs of his transition were the fading scars under his pecs, a by-product of the "top surgery" he had at nineteen, during a gap year between high school and college. That was in 2015, which, some have argued, was the year transgender identity went mainstream. In 2015 Caitlyn Jenner appeared on the cover of *Vanity Fair*, the year that *People* named Laverne Cox one of the world's most beautiful women, the year that *I Am Jazz*, a reality show about a trans fifteen-year-old, debuted on TLC and the Amazon hit *Transparent* entered its second season. The ban against openly transgender individuals serving in the military also fell that year. Nine states pushed back with "bathroom bills" that would restrict access to those whose gender matched their designated natal sex; none passed. Trans rights had suddenly become the cutting edge of civil rights, a litmus test for progressivism, and teens like Devon—whose parents are liberal, educated, and affluent—were in an ideal position to benefit. He was not denounced or disbelieved when he came out, not cast into the street, not raped or murdered, in what has become a rising epidemic of violence against transgender Americans (especially those of color) as their visibility increases. As with Zane, Devon's family neither hesitated nor wavered in their support, whether that meant navigating school regulations or paying for a bilateral mastectomy and hormone therapy. Even so, the journey to define his masculinity and sexuality was not easy.

Three years earlier, as a high school junior who still thought of himself as female, Devon came out as a lesbian to his friends and his family ("I was never worried that my parents wouldn't accept me," he said). But he stayed in the closet on his swim team, afraid that he'd be subjected to further harassment. He also began dating one of his best friends off and on but recoiled whenever she tried to touch him. Kissing was okay—everyone kisses, regardless of gender, so it wasn't "complicated." But anything else made him uncomfortable. "I was so locked up in terms of my body. In order to feel good you have to relax, and I would just lie there, rigid. I hated her touching my breasts—which makes sense, since I hated my breasts—but I thought I *should* like it, so I would just be silent. And genital stuff: it's not like I wanted a penis, but I was so disconnected from my body and my womanhood, and so much of sex is about being connected to your gender."

He'd expected that coming out would resolve the growing estrangement from his body, but it didn't. Instead, he was more miserable. He continued to excel at school and swimming—he was nothing if not a perfectionist—but increasingly began to restrict his diet: bingeing, purging, oscillating between anorexic and bulimic behavior. More than once, he considered suicide; obviously, he was depressed. His therapist recommended that he enter a residential program for women, but Devon wanted to walk across the stage for high school graduation; his parents agreed to let him if he promised to start treatment immediately afterward, deferring college for a year.

Epiphanies can happen in the most mundane moments. In Devon's case, about a month into his program, he was riding in a van with some of the other patients. He'd been spacing out while they talked about "girlie stuff" and happened to glance down at his lap. He was dressed in the baggy denim that he called his "man jeans." His therapist had recently asked why he'd worn boys' clothes through his childhood; Devon replied that he'd never thought much about it. Now, though, it occurred to him that the discomfort he felt in his body, an agony so acute it had landed him in rehab, was not actually about being "fat." Not really. It was more that. . . . He disliked how that body fit his pants. "And I realized," he recalled, "that I would actually feel a hell of a lot better if my body was thirty pounds heavier but filled my clothes in what I found a masculine way than if I lost another twenty pounds.

"And I thought, *What if I don't like my body because I'm not a girl?*"

For a couple of months, Devon turned that idea over in his mind, recognizing and resisting it, knowing it was true but not wanting it to be so. He worried about telling his parents. He worried about swimming. He worried about what he called "losing lesbians" (when some members of a lesbian empowerment board posted that trans men were tools of the patriarchy, "that fucked me up for a couple of weeks," Devon recalled). Finally, he enrolled in a workshop on gender where he could meet some trans people, "and this kid walks in," Devon said, "about thirteen years old, and I'm double-taking because I can see myself in him. The way he walks that's a little more feminine than you'd expect. His chest was bound, which I used to do. He wore cargo pants, just like I would. But he had facial hair and this low voice. And when he started talking about being a trans person, I started bawling and I couldn't stop. Every other line felt like, 'This is me! This is me!' It tore my world apart. And it gave me the spark to start a new one."

In the fall, Devon began hormone therapy. He wanted to have top surgery immediately, but his parents, who he said have supported him fully, were nonetheless reluctant for him to rush into having his breasts removed (most transgender people, whether male or female, do not transition surgically, either out of choice or due to lack of funds and discrimination by health insurers). Devon's father accompanied him to a facility in Florida that specialized in female-to-male transition—the staff told them it was the first time that a patient came with a male rather than a female parent. Devon wept with relief when he woke from the surgery. "It was a huge weight off my chest—ha ha," he quipped. "I was passing before, but now I pass all the time." His only real concern was swimming—according to NCAA rules, the hormones made him ineligible to swim as a woman. It took some negotiation and advocacy by his parents, but by the time he started school the following fall, he'd been accepted on the men's team.

Part of what intrigued me about transgender boys, and why I wanted to be sure to include them in this book, was that they provided a unique perspective on masculinity. Devon, much like cisgender gay boys, talked about learning to walk, stand, and sit "like a guy"; take up more physical space than women; use fewer hand gestures; plant his feet and squelch the tendency to give nonverbal conversational prompts (women tend to nod or repeat an encouraging "uh-huh"—men do not); and master the art of the bro shake (there are YouTube guides for that). Unlike cisgender gay boys, however, he also spent years going through the world in a female-presenting body—a "hot" one, at that—with the inevitable exposure to critique, harassment, the ever-present threat of assault. Every trans boy I met commented on how it felt to cast off that hypervisibility. Long after his surgery, Devon would tense on the street when he heard a catcall, only to realize it was aimed at someone else. Hutch, another trans boy who was Korean American, remarked that as an Asian woman he felt on edge from puberty onward: fetishized, harassed on the street and in school hallways. During a gap year abroad, he was raped. "Now that I'm perceived as male, I don't experience any of that," he said. "My gender is still racialized, because Asian men are typically viewed as effeminate and less desirable, but even so, as a man, you're able to go through life and not fear that your body will be violated—I mean, yes, there are male survivors, but it's less commonplace, less incessant. That's one of the starkest contrasts.

"The other thing," Hutch added, "is because of my experience of being trans and being a survivor of assault, I think I'm much more aware of the importance of consent in any kind of sexual encounter, especially now that I have male privilege. I wish sometimes I could have a conversation with men about the way they treat women. I'd still like to know how they justify it."

Once he was on the men's college swim team, Devon suddenly had that opportunity. Even at his elite, liberal university, the "locker room banter" could sometimes be hard to hear, and like other boys I met, he wasn't sure how to respond to it. Maybe the fact that he was trans made his position more precarious, but maybe not: recall how Cole described his quandary on the crew team in high school, or Ethan, who transferred colleges rather than confront other lacrosse players. Early in his first semester, Devon, too, wanted to quit. He called his parents during a team event that happened to be held a few hours away from their home and asked them to pick him up. "Joining the team was a terrible decision," he told them.

They urged him to hang in there a little longer. "Just try," they said.

He hung up and wandered outside. The team's captain followed him, asking what was wrong. Devon took a deep breath and told him: he was upset by the sexist jokes they were telling. The other boy seemed genuinely confused. "Someone just told one!" Devon said. "He said, 'What's the difference between jam and jelly? I can't jelly my dick down your throat!' That's an *awful* joke!"

"But," the other boy protested, "it's just supposed to be funny!"

So Devon described what it had been like living in a female body, explained the concept of "rape culture," told the boy how those so-called jokes perpetuate it. "You have to understand," Devon said. "I've been the object of those comments, and it hurts to hear you talk that way."

To his credit, the other boy listened and apologized. "But you know, we didn't come to this from the same perspective as you, so we don't think about it," he added.

Devon ended up talking with a few other boys, who were concerned when they saw him leave, and, more critically, to the captain's father. "He told me, 'Devon, you can make a change. I've already seen it happening. I believe your presence will make a difference, but it can't make a difference unless you stay.' And I believed him." Two years later, Devon thinks it's working. Not always—sometimes he even finds himself about to say something crass or sexist to try to fit in—but it's definitely better. "Some of those guys who made those original comments, they have also been unimaginably kind to me in ways that have blown my mind. I realized that if they could be kind in a moment, they could be kind period. They weren't saying stupid things because they were stupid—they were saying them because that's how they were socialized, and that socialization shouldn't override my trust in their capability of being good guys.

"I know for a fact several of them used to have different opinions about trans and gay people and my being on the team has changed those opinions. I hope that's extended to the way they talk about women, too, but, then, it's not like I'm a messiah who's come down to change boy culture."

Of course he's not, nor should he have to be. What's more, trans boys can be just as misogynist as cis boys, something that frustrates Devon. "I think, *Are you fucking kidding me? I know people said that about you when you presented differently. Are you forgetting where you came from?* And oftentimes they are."

The Grindr Grind

Zane and I sat one frigid Midwestern afternoon at a hipster café a few miles from campus, one of those places where the barista sports a French handlebar mustache and pours a five-dollar cup of Kyoto-style coffee. Zane wore silver-coated jeans and a black turtleneck; under his John Lennon sunglasses, his eyes were darkly lined, and his brows, skimmed by artfully mussed bangs, were plucked thin.

That tortured syntax Zane used when coming out to his mom—"I'm not straight"—was actually more accurate than he initially realized. Like many "not straight" boys I met, Zane referred to himself as "gay," even though the word felt dated to him, as constricting and antiquated as "homosexual" to a previous generation. "The way I think of 'gay' at my school is the white boy who takes on the hetero ideals instead of questioning or undermining them," he said. "He presents as 'masc' and has an aversion to the more 'twinky' boys and feels more entitled than ever, with the sexuality difference ameliorated, to blend in and embody the privilege of his straight brothers. So it's strategic and political to call myself 'queer.' It describes everything about me. I like the ambiguity. It feels safe."

"Queer" for Zane meant that it was all up for grabs, open to question: masculinity, sexuality, monogamy, virginity. "I've been trying to move away from that narrative that as a gay man, 'losing my virginity' has to mean penetrative anal sex," Zane said, though, even so, that was how he ultimately defined it for himself. The spring of his freshman year, he met a student on Tinder from a nearby college whom he described as "the stereotypical 'femme' boy who loved Lady Gaga, which is what a baby gay like me thought 'gay' was." They traded messages for about a week—an eternity for gay men in the swipe app era—then progressed to text and Snapchat. Once Zane was reasonably sure the guy wasn't an axe murderer, he invited him to his dorm for, depending on how things went, a get-to-know-you date or a hookup. They had a glass of wine, chatted for a while, then the guy suggested that, although he preferred to "bottom" in anal sex, he could "top" if Zane wanted, so that his first time would be with someone who cared about his well-being. After some further discussion, Zane decided to go for it.

I was consistently struck by how much more willing and able queer boys were than straight ones to negotiate sexual consent. Their process, it seemed to me, could be held up as a model. Dan Savage, author of the syndicated advice column "Savage Love" and founder of the It Gets Better Project, agreed. He referred to "the four magic words" gay men will use during a sexual encounter: *What are you into?* "When two men go to bed together, 'yes' is the *beginning* of the whole conversation," he told me. "Who is going to do what with whom cannot be assumed. And when that question is asked at the moment of consent—'Yes, we're going to have sex. *What are you into*?'—you are empowered to rule anything in and rule anything out." That's a very different approach from that of straight boys, who usually aim for a simple "yes" or "no" to options they define, such as "Do you want to go down on me?" or "Should we have sex?" *What are you into?* is the kind of open-ended question that invites true collaboration and mutuality, not to mention, Savage pointed out, a broader definition of "sex." "It shocks straight people when I say the answer to 'What are you into?' for gay men is often, 'I'm not into anal.' Straight people project their default setting of sex-as-penetration onto a gay encounter, when often what you get is oral sex, or mutual masturbation, or rolling around, or fantasy play. And we do not regard those other things as consolation prizes; that's *sex.* Can you imagine if a straight guy asked a [straight] woman, 'What are you into?' And she said, 'I'm not into vaginal intercourse?' His head would explode." Of course, gay men may still end up in situations where there is poor communication, where they feel pressured, are taken advantage of, or are harmed. But, as Savage put it, "What that means is that 'gay' is a conversation. And that's what straight people should take away from gay people besides sit-ups and brunch—how to have that conversation about sex."

Zane never saw his first partner again. That was okay; the sex was fine, but they didn't click intellectually or romantically. In the meantime, at least at first, using the dating apps felt liberating: not only could Zane indulge sexually, but he could effortlessly expand past the borders of his own campus, whose population held little appeal. "The apps have allowed me to meet kids who are likely to be interested in what I'm interested in, or even people who are completely different, because that's interesting, too," he said.

The straight boys I met also used swipe apps, though less frequently and with marginal success. For them, the yield-to-effort ratio was low: there was a lot of matching with girls and a fair amount of chatting ("Hey!" "'Sup?"), but few actual meet-ups. What's more, although some women sought sex on the apps, more made it clear they were searching for dates or relationships, limiting the possibilities for no-strings hookups. Tinder, Bumble, and the rest seemed more a source of entertainment, akin to a video game, than of real-life prospects.

No question, dating apps increase the supply of potential partners, but everyone I interviewed who used them—male or female and regardless of sexuality—complained about their superficiality, their reductiveness, and the corrosive effect of continual, arbitrary rejection. "I've matched with guys who seem really neat," Zane said, "and then I say, 'Hey, how are you?' and never get a response. It doesn't seem like a big deal, but if it happens over and over again, you start questioning, 'What's wrong with me?' I have to calm myself down and say, 'There's nothing wrong with you. These guys don't even know you. They don't get to make a judgment.' But it can still be emotionally rough."

Over the next two years, Zane hooked up with around forty-five men, mostly through Grindr, which is to Tinder what a booty call is to a candlelit dinner and a movie. Profiles feature just one photo that fills the entire first screen: typically a torso shot, often headless, sometimes clad only in undies or a bathing suit. Far below that users can add a few brief descriptors. "Everyone is reduced to their bodies in Grindr culture," Zane said, and the judgment can be cruel. Someone might block you because you haven't shaved off enough body hair or because you've shaved off too much. Or for being too thin, or too fat, or too femme. Or for refusing to send an ass shot or a dick pic in response to the first "Hey!" (one gay boy said to me, "It's kind of like a resumé," perhaps illuminating one reason straight guys, too, send unsolicited dick pics), or because they don't like the ones you did send. What's more, this supposedly safe space for queer men became notorious for racism toward Asian Americans and African Americans: the company's 2018 #KindrGrindr initiative took aim at the common profile practice of including phrases such as "whites only," "not into Asians," or "no blacks."

Zane was a legal adult—eighteen when he downloaded the app. Other boys I met were not. Whether they were openly gay with limited options or closeted, Grindr had become their outlet—a safe distance from that neutered public expression of gay-best-friend sexuality—for exploring actual sex. Elliot, a high school junior at a private school in Baltimore, would tell his mom (who didn't know he was gay) that he was going out with friends, then drive to meet a stranger for a quickie; his partners were generally in their twenties but occasionally decades older. Elliot claimed to be nineteen on the apps. It wasn't what he wanted, he said, but seemed like his only option. "All of my straight friends have had relationships. I'd like a boyfriend, too, but this is what is presented to me." Neither Elliot nor the other high school boys I talked with disclosed their activity to friends and certainly not to parents, so no one knew where they were or what they were doing. "I always get nervous," Elliot admitted, "because I don't know whose house I'm going to be in. How is he going to talk? How is he going to smell? What is his place going to look like? Am I going to die right now?" Recently, he had met a man at the Fairmont hotel in Washington, DC, a locale he considered riskier than someone's house. I suggested that if things went wrong in a hotel at least someone might hear him scream; his partner would have had to put down a credit card for the room, too. At the risk of sounding overly dramatic, in a private home he could be chopped up into little pieces and buried in the backyard and no one would be the wiser. "You're right," he responded. "I never thought of that."

If Elliot had been a teenage girl using swipe apps to have sex with random older men, I would likely have felt compelled to report the behavior to an adult. So why, I asked him, was this any different? "I don't think it is, really," he admitted. "It's not ideal."

The trouble is, there aren't a lot of outlets for gay teens to form romantic attachments or explore sexuality in an age-appropriate way. Even if they are out, other boys around them may not be, limiting their possibilities. "If choosing between an inappropriate relationship and celibacy," Dan Savage said, "a horny teen will pick 'inappropriate' every time. Straight teens have more options and more support. Providing no channel or outlet for gay teenagers, telling them that they have to be eunuchs or aren't allowed to act on desire sets those kids up for disaster. It drives them onto Grindr, where good and decent guys will block them, leaving the shitty guys, or the twenty-four-year-olds who really do think you're eighteen."

It is perhaps not surprising to hear that straight parents of queer boys feel uncomfortable and ill-equipped to broach the topic of sex with their sons; parents of straight boys are flustered by such conversations, and at least in that case they understand the mechanics. Talking about sex with a gay child means directly acknowledging that physical intimacy isn't solely about reproduction, yet, in truth, that is exactly what teenagers of *all* sexualities need: ongoing discussion and education that addresses pleasure, mutuality, safety, love, intimacy, and self-discovery. They need to understand the potential to be either the perpetrator or the victim of intimate partner violence and sexual assault. They need to have agency over their bodies. And for all of them, adult denial puts them at risk of physical and emotional trauma.

Zane did find what he considers his first real boyfriend, a student at a nearby college, through Tinder, though the relationship took an ugly turn after only a couple of months: the other guy was secretly filming Zane, using their relationship as material for a documentary he was making about his exploits on swipe apps. Other experiences were even more disheartening. There was that "Republican gentleman," who lived in a swank apartment, had a boyfriend, and treated Zane like "a common street whore," expecting him to perform oral sex without reciprocation (this was another area where straight girls, who tend to presume sexual inequity, might take note; the gay boys I met found encounters that consistently went one way to be degrading and wouldn't put up with them for long). Or the guy who showed up at Zane's apartment, sweaty after a run, declaring that he had exactly eight minutes before he needed to leave to meet some friends.

Then there was that night the spring of his sophomore year when he decided to meet a guy he'd been talking to on Grindr. In retrospect, Zane said, he probably should've turned back when his Uber collided with a bicyclist: Wasn't that an omen? Plus, he'd been smoking weed earlier in the evening; he wasn't really high, but it had made him anxious. It was clear, as soon as he entered the guy's apartment, that something was off: the man had Googled Zane and would randomly bring things up about him in a way Zane found "creepy." Zane sat on the couch, and the man flipped on the TV. Then, without any warning, he shoved Zane against the cushions and jammed his tongue down Zane's throat. Zane tried to pull away, but the other guy, who was taller and heavier, pinned him. "I could not move," Zane recalled. "I felt like I was using all my strength to get him off and he wasn't budging." Finally, he wriggled away and said he needed a moment. "I sat on the couch in silence for what felt like forever, but was maybe ten minutes," Zane said. He had, since his freshman year, been active in his school's anti–sexual assault efforts. "My mind-set at that time had been that in the larger discussion we needed to move some of the conversation away from these low-level situations. We weren't paying enough attention to the more violent scenarios, the kind that women of color face, the kind that are legally actionable. And I still think about that. But I didn't really understand how coercive words and actions could be." When the man asked if Zane wanted to go into the bedroom, he went. "I just couldn't say no for some reason." He performed oral sex on the man, and then the man reciprocated, though Zane just lay there, wondering why he was allowing it. Afterward, all he wanted was to go home. But when he tried to leave, the guy grabbed him by the wrist, pulled hard, and said, "No! Stay here awhile!" It felt more like a command than a request. When Zane tried to get up again, the man lifted him off the ground and threw him back onto the bed. "I never imagined that something like that could happen to me," Zane said. "I'm not going to call it assault, but I'm not going to *not* call it assault."

Zane tried to forget about the experience, but over the next few months it intruded into his thoughts, especially during subsequent hookups. By summer he would have "a fucking panic attack" if someone touched his back a certain way or kissed him too roughly. "It was making my sex life worse because these things were affecting me, but at the same time they actually *were* fucked up. It made me realize that in none of the sex I had were people paying attention to *me*. What happened that night was just an exaggerated form of a lot of other things." By the time he went back to school for his junior year, he needed a break: he deleted the apps from his phone.

At least, he did for a while. That winter, just before Christmas, Zane texted me: *Thought you should be the first to know that I had the HEALTHIEST sex in ten months last night. It gave me thoughts on what a "good" hookup is.* It was, he said, all about connection. He'd met a guy through Grindr that he'd also talked to on Tinder a year before. "The Grindr norm is to exchange a couple of texts—'Hey!' 'Hey!' 'How are you?' 'Good.'—then you send nudes and people make a decision about whether they're going to message back. But we didn't exchange pictures, and we had this nice conversation, so it felt like there was a little familiarity." They met and hung out for a while at the guy's apartment, chatted about school, had a couple of glasses of wine. After a few episodes of *RuPaul's Drag Race*, the other guy asked, "Do you want to do something else?" which was exactly the kind of open-ended question that put Zane at ease. They went into the bedroom, listened to a few cuts of Bowie on vinyl, and then they began to make out. They took it slowly, gradually, making it clear that they both wanted to be there. The other guy listened to Zane, cared about his comfort and pleasure. When they were through, they lay in bed talking for a while, then the other man walked Zane to the door. They Snapchatted for a couple of days, but winter break was coming up and Zane was spending the next semester abroad, so he didn't pursue it. Still, the encounter had felt . . . personal. Zane felt seen during sex in a way that he hadn't for ages, more rather than less humanized by the experience. "For so many gay men, the thrill really is the anonymous, quick sex," Zane said. "And that's fine. I don't want to sound like I don't respect what other people want, but I think that as a community, we've gotten to a place where we don't value intimacy or connection. It's framed as liberating and sex-positive and people finding out their tastes, and I think a little of that is true, but it's also another ground where violence can occur."

I asked Zane whether he could imagine an alternative to that for himself, something that would still feel personally or politically "queer." "For me as a gay man, demanding more intimacy in some of these relationships actually feels subversive, even though it's more heteronormative," he said. "And maybe someday I will be in a monogamous relationship, and maybe what's 'queer' will be more about household chores or our work life, or the people we're friends with or where we live or how we dress. I don't know.

"I think," Zane added, "that I'm learning to acknowledge my own vulnerability. I've never done that before. And I'm making myself demand more of other men, too."

When Is a Man a Man?

Devon, the trans swimmer, didn't hook up much; his teammates did. And that affected his ego exactly like other boys'. "I worry that guys on my team see me as less attractive and desired because I don't leave every party with a girl," he told me. "I do recognize that isn't my manhood, and I can separate myself from it, but whether you call it 'toxic masculinity' or something else, being invested in the idea of promiscuity is still part of growing up male." His freshman year, Devon admitted, he even used an ex-girlfriend, Haley, to curry status.

"She's so hot, dude!" a teammate said when Devon showed him her picture. "How'd you get her?"

"I never *wanted* to treat women like objects," Devon said, "but I have to admit there was a piece of me that liked that approval. I worried that they wouldn't think I was muscular enough or fit enough or that my voice was low enough, but at least I was man enough for this beautiful girl to love me." As for Haley, who had previously dated football players, Devon suspected she'd been exploring her sexuality with him, that he was "a bit of a puzzle piece." They're still friends, but, in retrospect, he realized that regardless of her motivations, he wasn't yet fully ready to have a sexual partner: his body was changing—his hips slimming, his shoulders broadening—and he still disliked being touched. "She would pressure me to let her do things to me and, this is really fucked up, but she would say how she was really good at giving blow jobs. It would make me feel inadequate, like I should have a penis. I don't think she meant it that way. It's just . . . she knew what she was doing with those parts and she didn't know what to do with mine."

Haley also suggested that, to better understand sex and masculinity, Devon should watch porn. Huge mistake. "I don't blame her. We didn't know any better, but I definitely, definitely did not learn anything positive from that." Before watching porn, Devon said, he never had a picture in his head of what sex—any sort of sex—looked like. "What I'd called 'sex' when I was presenting as a woman was what I'd done with a girl in high school: we made out and touched each other in places that felt nice. I didn't have an image or a judgment. It was sloppy, but we were figuring it out together. Once you watch porn, there is a clear image of what is normal and desired and what sex should be. And I'm wondering: Is *that* what I'm supposed to look like? Is *that* what I'm supposed to do? Is *that* what masculinity means? Those images are damaging to a trans person who will never have the body parts, but I think they're actually equally damaging to people who have the parts because those ideas aren't realistic for anyone."

After Haley and he broke up, Devon tried to join in the campus hookup culture, but he grappled with the ethics of disclosure. "It's like, do I tell a girl after I get drunk?" he said. "Before? Do I keep my clothes on so she won't know? If I don't tell her, is that wrong? If I do, will she yell at me?" The potential for disaster felt too high. Devon tried downloading Tinder, but, again, hesitated over how to list his gender. As male? As trans? As trans male? He eventually settled on "trans" (some other trans boys I met preferred "male"), and although he matched with a few girls and went on a couple of dates, nothing panned out. At one point, he even tried hooking up with another trans guy, thinking they might understand one another's bodies. "The sex made me feel violated," Devon said. "I stripped my bed right away after he left and washed the sheets. It wasn't that he did anything wrong, but it confirmed for me that I wasn't interested in men, whether or not they had a penis."

I admit, I sometimes struggle to understand a new generation's approach to gender. Does it challenge or reinforce convention when we consider certain attitudes, qualities, ways of presenting to be "masculine" or "feminine" regardless of a person's anatomy? Must a boy in a skirt be seen as possibly gay or female? Must a girl with a crew cut and baggy jeans be thought of as mannish? Also, while some nonnormative young people try to explode the binary by identifying as fluid, others, like Devon or Hutch, identify firmly and fully as male. Nor did that identification predict sexuality. Hutch, for one, enjoyed vaginal penetration and felt most sexually attracted to what he referred to as "people with penises," though more romantically toward those with vulvas. It's confusing, but then, maybe that's the point. "When you have a shift, even a small shift in the lived experience of gender around trans bodies, there's a domino effect on heteronormativity," pointed out Jack Halberstam, a professor of English and gender studies at Columbia University who writes about female masculinity and trans issues. "So if a woman flirts with someone she finds out later is a trans man, it might inadvertently change her understanding of what it means to be a man whom she finds attractive. And that creates the potential for more positive, surprising, pleasurable interactions between people."

As with Zane, Devon felt a previously unimaginable level of support from friends, family, and community. Even his grandmother took his transition in stride. They could, and would, accept him as a boy—but, unsurprisingly, no one talked about what that meant in bed. How do you find pleasure in the body with which you've been so at odds, and specifically from the parts that may be fundamental to that disaffection? "I didn't feel *hopeless* about it," Devon said. "But I didn't have a very high level of confidence that I would ever enjoy sex for myself."

Then, midway through college, he met Mia. The first time they were naked together, she looked at Devon, really took him in, and clearly appreciated what she saw. "I thought, *God, that is a beautiful man!*" she told him later. When Devon said he didn't want to be touched, Mia said, "That's okay, I don't have to touch you. But I'm going to make you *want* me to." And she did.

Their relationship has not been a fairy tale: it has been turbulent, inconsistent, sometimes tense. But sexually, it has made Devon feel whole. "I have felt like she welcomed me into my own body," he said. "She's so reverent about it and that is so healing for me. She says, 'I love your parts, I love doing what I do to you, and you are male to me.' Through that, I've learned to feel that way as well. She's taught me that just because my body isn't classically male, that doesn't mean that *I'm* not male."

Devon and Mia have been able to improvise, to build a sexual relationship that feels communicative, creative, and authentic to the two of them. Again, that's not by any means always the case for trans people, many of whom stick with what Devon called "touch-me-not," but it is a potential benefit of being forced off the classic script. Devon didn't talk about any of that with his teammates—he worried discussing sex would highlight their differences—but he did consider what it might mean if more cis guys could be open about what was really going on for them behind closed doors. "Not this whole idea of"—Devon dropped his voice into "bro" mode—"'Dude, we fucked all night and it was great. And she gave me head and she swallowed.' I wish they would talk about sex more like girls do, or at least some of the ones I know: they talk about what felt good and what was fun and what the person did right. It's not so much about conquest. If I were to talk to other guys, or anyone, it would be about the idea that there are so many ways that we can interact sexually. Queer people are willing to step outside the boundaries because they have to or because they want to, and you find a lot of cool things about that—and it's fun."

Listening to Devon, I recalled something else that Jack Halberstam, the Columbia professor, had said: trans people may or may not personally challenge gender norms, but in the best-case scenario, their increased visibility would not only make the world a safer place for them but, by encouraging all of us to ponder our relationship to gender, race, and entitlement, could ultimately topple oppressive systems of power and sexuality.

"I do think it's important for everyone to question gender," Devon said, "because it helps you understand who you really are and what you believe. I don't mean to say that my teammates or other male friends have to question it in the sense of doubting whether they're straight, cis guys, but thinking about 'What does it *mean* to me to be a straight cis guy?' Making that visible. I am required to answer 'What does it mean to be transgender?' because of the way the world works. But what does it really mean to be straight? What does it mean to be cis? There's a choking off of identity that might be remedied by people just having more space to explore themselves."

I talked to Devon one last time shortly after graduation. "This year, I mentioned to a guy on my team that I had been afraid of talking about my sex life with him because I didn't want him to see me differently. He interrupted and said, 'Devon, even if you stripped down right now in front of me, I wouldn't see you differently.' That was a healing moment as well and indicative of the culture that had developed on the team. It's not like kumbaya-land, where we're all sitting around telling our secrets, but people feel more comfortable with their close friends and there was a lot more sex positivity and openness. And . . . I don't think that was an accident."

CHAPTER 5

Heads You Lose, Tails I Win

Boys of Color in a White World

Darkness was falling on a late November afternoon as I crossed the campus of a large state university in the Midwest. I shivered against the chill, shoving my hands deep into my pockets, wishing I had remembered gloves; I spent my childhood in Minnesota, but three decades in California had made me soft. The students I passed looked like those I'd seen at a dozen other campuses: girls in black leggings and North Face jackets, their hair hanging loose under knit winter hats; guys in jeans, duck boots, and (despite the cold) school hoodies. Normally, I would've paid little attention to them, and even less attention to their race, but today I was hyperaware: white, white, white, white, and white. Every student I passed was white. The girl who politely let me into her dorm without asking why I was there or whom I was visiting was also white; I'm sure she assumed that a middle-aged lady like myself—again, white—was no threat to anyone. The students I was meeting weren't extended that same trust.

Xavier and Emmett, both first-semester freshmen, were African American men, the most underrepresented group on campus, something they were reminded of regularly. More than once, when Emmett had shown up at a friend's dorm, or walked across campus late at night, he had been stopped by security guards, demanding proof that he was a student. Even when he showed his ID, they seemed skeptical.

Higher education is not a level playing field, not in the classroom and not in students' social lives. Only 40 percent of African American men graduate from predominantly white four-year colleges within six years of matriculating. Financial pressure, isolation, and the presumption that they should adapt to fit the culture rather than that the culture should expand to include them, as well as flat-out racism (both tacit and overt), take their toll. Addressing all of that is, obviously, beyond the scope of this book, although I will not ignore it. Nor do I purport in these pages to represent the experience of all-minority communities; my interest was in how the masculinity our culture sees as "neutral"—essentially a white, straight version—affects young men regardless of race, sexuality, or gender identity. So the guys of color I spent time with were a select, circumscribed group, the classmates of the other boys I interviewed: educated in majority-white high schools (although most were from lower-income neighborhoods compared to their peers') and attending predominantly white colleges—the putative poster children for upward mobility, canaries in the coal mine of an egalitarian dream. They often arrived on campus with a different mind-set about partying and hooking up from their peers. White students I've interviewed, both guys and girls, would talk about the "college experience" as a time to "go crazy," almost like a four-year trip to Vegas. At first I found that befuddling: my own "college experience" was more about self-knowledge than self-obliteration. It turns out, according to sociologists Elizabeth Armstrong and Laura T. Hamilton, universities have themselves conspired in that shift. As financial challenges have mounted—especially at the kind of moderately selective, flagship state schools attended by Xavier and Emmett—administrators have come under pressure to lure wealthier, out-of-state students who can afford their higher fees. So they have actively rebranded college life as four years of "fun," introducing an array of low-pressure majors for those who want to spend (at least) as much time drinking as studying.

In *Paying for the Party*, their five-year study of students at one such university, Armstrong and Hamilton found that all that "fun" was fine for those more affluent, socialite types: they may have ended up with a middling education, but their families' connections and resources set them up for postcollege success regardless. The more serious, professionally minded students also fared well—if they had wealthier, educated parents who were savvy about the system and intervened early to keep their children attentive to their goals. Only the less privileged students truly suffered under the university's new priorities. It didn't matter whether they were academically driven, committed to economic mobility and professional advancement: they received so little support that all of them either transferred to less prestigious institutions or dropped out entirely. Meanwhile, a sweeping 2018 study tracing the lives of millions of children found that even when black and white boys are raised next to one another in the same wealthy neighborhoods by parents earning similar incomes, as adults the African Americans earn less, and most become poor; there is no such differential between girls.

The black guys I met talked about those pressures and more as they navigated complex educational experiences. They discussed feeling at once highly visible and invisible on campus, perpetually watched yet also unseen. They simultaneously acknowledged and minimized the psychic cost, the emotional energy it took to weather small but continual slights in the classroom, in their dorm rooms, strolling across campus, at parties: the constant, low-level message that they did not belong. And although they expressed gratitude for the educational opportunities that most white students I met (male and female) took for granted, that was accompanied by a sense of precariousness: a persistent anxiety about how easily their privilege could disappear.

The Coolest Guy in the Room

When he was twelve years old, Xavier was offered a full scholarship to an exclusive private school, a golden ticket out of his low-income neighborhood in an East Coast city and into a comfortable future. Just before the semester began, his mom, a school bus driver, sat him down in front of YouTube to watch the infamous, 1991 citizen-journalism video of Rodney King being brutalized by Los Angeles police. "I'm not showing you this to scare you," she said, "but to show you what you're about to walk into."

Needless to say, that wasn't how the parents of his white classmates were preparing their sons for middle school.

Looking back, Xavier feels it was the right thing to do, even though it frightened him. "You've got to have that innocence broken at a young age," he told me. "Especially when you're going to be around people who are so powerful and so rich."

I first met Xavier when he was a senior in high school, one of a group of black guys I interviewed from his class. Broad-shouldered with a small soul patch and warm eyes, he had a kind of stillness about him, a melancholy, perhaps the result of losing a parent too soon: his dad had died of cancer when Xavier was sixteen. He wanted to be sure that I understood that his father had been in his life—maybe to offset any assumption that black children are all raised by single mothers—although he couldn't remember much about the older man before his illness. "My grandfather taught me what a man is supposed to be," Xavier said. "Just seeing how he supports my grandmother. He taught me to care and take care of people, that kind of selfless love."

Xavier was excited to start his new middle school, looking forward to the academic challenge and making friends. But fitting in wasn't easy. He'd gone through puberty early, so he was taller and heavier than most of his classmates, not to mention much darker-skinned: he was also the only boy of color accepted to his grade that year. "I tried to be friendly," he said, "but there was no real connection there. And I was kind of demonized as this big, black guy who acted sort of funny because I didn't grow up in a white neighborhood. So I felt very secluded. I didn't really speak a lot. I didn't feel like I had any real friends. It was rough."

He stuck it out anyway, though perhaps against expectations. He recalled running into one of his middle school English teachers several years later on campus. The man seemed surprised to see him.

"Oh, you're still here?" he said.

"I was like, 'Where did you think I was going?'" Xavier remembered responding. "And he said, 'I didn't think you were going to make it.'"

In that rarefied world, Xavier felt continuously scrutinized and *judged* not as an individual but as representative of his ethnic group. As was the case for the gay and trans boys I met—or any boy who deviated from the norm of cis, straight, white masculinity—Xavier was conscious of everything he did, everything he said:

how he walked, how he spoke, how he dressed. "I carry myself in a very professional and dynamic way in public spaces," he explained. "For myself and for my family, but also to prove . . . You know, 'He's not going to come in here wearing some ripped-up jeans' and this and that. It's like, 'Yes, I actually do have a pair of dress shoes.' It was almost like I needed to show other people who've subscribed to a lot of popular black culture that there's more to us than what they see in videos."

All that said, African American boys have been found to have an easier time than girls integrating socially into largely white schools. Both sexes are stereotyped by their peers, but the guys are admired, considered "cool," "athletic," "street smart"—like a hip-hop star fallen to earth—while the girls are labeled as "loud," "aggressive," or "ghetto." In Xavier's school, white girls had been only too happy the previous year to attend prom with black male classmates, but few girls of color were asked by boys of any ethnicity. "Which it shouldn't be, but that just shows how everything is," he said. The black girls weren't typically invited to white classmates' parties, either, whereas hosting the guys conferred instant street cred. "We have the coolest music, the coolest dialect," Xavier said. "Black culture basically *is* the culture, and that's the ideal of what a lot of white males want to be. And I guess I'm the closest they're going to get to that."

Maybe so, yet the version of black culture marketed to white kids has little relationship to who the young men I met actually were. Xavier, for instance, was an Eagle Scout and a summer camp counselor and represented his school at a national diversity conference. His classmate Aidan, seventeen, who was slim, with a low fade Afro and a spray of acne across one cheek, told me he preferred old-school Lauryn Hill to "anything that uses the word 'bitch' in relation to a female"; quoted Frederick Douglass and

Cornel West; and was active in his school's sexual consent advocacy group. The commercialized blackness their classmates consumed, by contrast, often promoted hypermasculinity, violence, and degradation of women, presenting those as proof of authentic manhood. It's all laid out perfectly in Lil Dicky's 2018 novelty hit "Freaky Friday," which, at this writing, had garnered over a half billion views on YouTube. The comic rapper, who is white, wakes up in bed with two scantily clad women, having magically been transformed (by an "inscrutable" Asian waiter at a Chinese restaurant) into R&B artist Chris Brown. He looks into his pants to discover his "dicky" is—*surprise!*—"lil" no more, and realizes that he is suddenly a spectacular dancer. Chris Brown, on the other hand, trapped in Lil Dicky's body, recoils from the sight of his diminished sex organ, but finds comfort in being able to wear the color blue (historically identified with the "Crips" gang) without being shot.

The video plays with how white guys see (and are conditioned to see) black masculinity: as simultaneously aspirational and threatening, what sociologist Michael Kimmel has called a defiant mix of "athletic prowess, aggression, and sexual predation." But adopting those trappings—through what they listen to or watch, in the posters on their bedroom walls, in how they talk, dress, walk, dance, in whom they idolize—is as likely to reinforce as to challenge stereotypes. Guys like Xavier and Aidan, who grew up in some of the poorest neighborhoods of their city, are both advantaged and constrained by the image projected onto them. "You have to be careful," Xavier said. "People will treat you like you're the coolest guy in the room and everything, but they'll also think you're the *dumbest* person in the room. So down the line, it won't really help you unless you show a different side of yourself. You basically always have to prove yourself."

Xavier hadn't been to a party since his sophomore year. "The idea of drinking in a situation where there's fifty drunk white people . . ." He shook his head. "I don't feel comfortable at all." He met his current girlfriend at a local diversity conference. They were friends for a year before things turned romantic. "I appreciate that as a woman of color, she can understand me in ways that are multidimensional without my having to explain anything," he said. "I just find more peace in it."

Xavier's classmate Emmett, on the other hand, who had been at the school since kindergarten (and would, by coincidence, end up at the same college as Xavier), was happy to trade on his cachet among white kids—in his own neighborhood, he never would have been seen as cool. Gangly, with heavy-lidded eyes, his hair styled in short twists, he often referred to himself as fundamentally "lazy": too lazy to work hard at school, too lazy to make the first move in a hookup, and certainly too lazy to transform a hookup into something more substantial. "Practically everything is a joke to me," Emmett said. "It's just how I was raised. I suppress all my emotions and play everything off as funny. Some girls try to cuff me or be girlfriend and boyfriend, but"—he held his palms up—"I'm like, 'No! No!' And usually, girls know what to expect if they're hooking up with me." At seventeen, he had never gone on a real date ("Maybe once or twice something where we say we're going to go get lunch and then we completely disregard that and go back to her crib"). Instead, girls from his school would hit him up on Snapchat, disconnected from the messiness and potential rejection of face-to-face contact. They'd talk and flirt, and by the time the weekend rolled around, the two of them—and everyone else in their social circles—knew it was on. Emmett had been with "a lot" of girls that way, he said, mainly at parties, after which he'd go back to hanging out with his friends, a small group of black guys who also attended his school. Most of the girls he'd hooked up with were white, partly because that's who was in his community, but also because, that way, he was less likely to get attached. Usually, he was stoned at the time. "Making out with someone when you're high is not enjoyable because you can't really move your mouth," he acknowledged. "But getting a blow job when you're high, that's like the best thing ever. So that's also part of it—if you're already set up with someone on Snapchat or whatever, you can skip all that other stuff."

When I asked whether he ever performed oral sex on a partner in return, Emmett actually gagged. "I guess it could be kind of disrespectful that I don't, but, I mean, it's, like, a lot to ask." He paused, ducking his head. "No, no, it's not a lot to ask. And I think if you're in a relationship, I think definitely. But, like at a party . . . I also had a bad experience the first time I went 'down there.' It was when I was a sophomore, and afterward I threw up. It was just terrible." At any rate, if he liked the girl, maybe he'd follow a hookup with a little Snap streak, but often he never checked in again and neither did they.

In many ways, then (with the possible exception of vomiting after oral sex), Emmett's hookups were like those of many guys his age who were into the party scene. But then our conversation started to turn. "You know the term 'designated driver,' right?" he asked. "My black friends basically have a 'designated watch-out person.' Like, I've seen white boys dragging a girl around a party—not dragging her by the hair or anything, but not allowing her to do anything or really talk to anyone. And that is misconduct. If we did that . . . Or when a white girl says, 'Back off,' and the guy doesn't? I've seen that plenty of times with white boys. But if that same thing were to happen with one of *my* friends—I mean, most of us would just back off when she asked, but if someone was really under the influence and just didn't know what was going on—we have someone there to watch out. To keep him from doing something he's going to regret or maybe doing something illegal, or even getting with someone that he wouldn't want to get with sober." That might seem a good idea for anyone, really (and some college frats and sororities designate "sober monitors" precisely for that reason), but black guys in white worlds knew that for them the stakes for crossing a line would be higher and the bar for unacceptable behavior lower. The qualities projected onto them by white peers—greater physicality, sexuality, a latent potential for violence—could quickly turn from desirable to dangerous, from cool to predatory. As one boy of color I spoke with said, "That's why my mom's going to drill it into my head: you can*not* trust white women."

Not the "Typical Asian Guy"

During high school, back in Northern California, Spencer didn't think much about race. The child of Korean immigrants, he went to a school in an affluent suburb with a sizable Asian American population, so he never felt unusual or different—his identity was grounded less in ethnicity than in being a baseball player, an artist, generally well liked. "My closest friends were mostly Asian," he said, "but I had a lot of friends in high school, and other groups were much more mixed." Things had changed since he started college in the Midwest. "I used to find it easier to talk to girls, hook up with girls, but in college it's . . . Honestly, I don't know. I just find it a lot more difficult."

Spencer, twenty, was sitting on an office chair in the bedroom of the off-campus apartment he shared with three other guys; I was on the bed, leaning against a cinder-block wall. He wore a backward baseball cap, baggy shorts (although it was December), and Top-Siders. He would describe himself—repeatedly—as short, though I hadn't noticed. I'm five seven and change, and when I stood to leave, the two of us saw eye to eye, which, while making him slightly under the average height of an American man, was hardly freakish. Plus, his broad shoulders, muscled neck, and athletic frame projected a much larger presence. Spencer was a member of a frat whose reputation when he joined was, he said, "kind of douchebag-ish: people who didn't care about other people and wanted to have a good time was the vibe." Personally, he felt that was unfair: he'd found his brothers to be "genuinely good guys" who seemed to legitimately care about him, a lot nicer than the other frats he'd rushed, which he described as "druggy" or "sketchy." He made a point, too, of telling me that twice in the past two years when complaints of sexual misconduct were lodged against members—both times for nonconsensual digital penetration of girls on the dance floor—the boys were immediately kicked out. On the other hand, that was still at least two cases of sexual misconduct in a relatively short time.

As was the case for many white boys I met, then, Spencer's social life revolved around his sports team and Greek life. On a typical weekend, assuming he wasn't in training, he'd pregame with his frat brothers, knocking back six or seven drinks: one of their favorite pastimes was making one another take a shot every time they died while playing the video game *League of Legends.* If the night's main event was at their own house, Spencer would keep going, playing beer pong, doing more shots with friends—downing maybe another eight or nine drinks. If they went to a party elsewhere, where his access to higher-shelf liquor might be limited, his total for the night would be lower, closer to ten. Either way, it was a lot of booze, especially for someone who claimed to hate the taste of alcohol. But Spencer didn't see the point of drinking if you weren't going to get fucked up, and the faster you threw them back, the easier they went down. He'd only blacked out twice, he said, and he tried to stop before he puked, though he couldn't always recognize when he passed that limit. "I'm just a lot more confident when I'm drunk," he explained. "All the positive sides of me come out. I'm very outgoing. I talk a lot. I guess I find myself more fun to be around.

"And, this probably doesn't help," he continued, "but I've also heard before that other people like me better when I'm drunk, which . . . When you hear someone say they like you better when you're on something that makes you not yourself, that kind of hits home. It kind of hurts, you know?"

Spencer had never had sex sober, though, in truth, he hadn't had sex much at all—two hookups during his three years of college, each time a drunken one-off—which was not so unusual, statistically, though he believed it was. He compared himself to a high school friend attending a Big Ten school, a guy who hadn't had his first kiss until two years after Spencer but was now on a "Tinder rampage." (Spencer's own profile picture, taken several years earlier, would seem to obviate that possibility: he looked about twelve. Other shots showed him clowning with friends, clearly wasted. His professed favorite pickup line was, "Are you Google? 'Cause you've got everything I've been searching for.")

Spencer lost his virginity during the first weeks of freshman year, the time when students "go crazy" with their new freedom, when the heaviest drinking and partying tend to occur—as well as the highest rates of assault. He had met the girl once before, briefly, at a pregame gathering. When he saw her again, a few days later at a frat party, he was about ten drinks in; he has no idea how much she'd consumed. They started grinding on the dance floor, then kissing, and somehow—Spencer doesn't really remember how—they ended up in his room. The whole episode would've felt awkward given that they were relative strangers, but he was too drunk to care, or, for that matter, to notice how the encounter was going for her. They barely spoke while it was happening. Or after. Or, really, before. Spencer was glad to have been able to unload his virginity, but beyond that, he said, "It wasn't too big of a deal. Obviously, I bragged about it to my friends because that's kind of what you do." The girl made no effort to get in touch again, so neither did he.

His second hookup, sophomore year, was during a "heaven and hell" theme party at his frat. Spencer was doing shots with his baseball pals when a girl with whom he'd occasionally Snapchatted started flirting with him. Again, grinding led to making out, and his inebriation emboldened him to ask the girl up to his room. Except he had "whiskey dick": he was too drunk to get an erection.

"Maybe we can do this another time," he offered apologetically.

"Yeah, totally," she agreed. She gave him her number, and they returned to the party, immediately separating. They did eventually go out on a date, but it turned out that, sober, they had little in common. Spencer invited her to another party, but lost interest when she started making out with someone else right in front of him. "If she wanted to hook up with me again, obviously I'd probably be down," he said, "but I would not pursue her romantically after that."

Both of the girls in those encounters were white. That was not a coincidence. "On one hand, I find white girls more attractive than Asians," Spencer said. "Then on top of that, I guess there's the byproduct, that 'Oh, he is special,' you know? 'He's not the typical Asian guy.'"

The politics of race, attraction, and love are complex. I've known women of color, especially Asian women, who wouldn't date a white guy before asking how many women of their race he'd previously been involved with (Jordan Peele played on that idea from the male perspective in his horror film *Get Out*, when the protagonist discovers a trove of photos showing his white girlfriend with her many prior black boyfriends, though she'd claimed he was her first). And who decides the magic number that separates open-mindedness from fetishization? Asian men hailed the 2017 indie film *The Big Sick*—a loosely autobiographical rom-com about the relationship between a Pakistani American man and a white woman—as a blow against cultural emasculation; South Asian women were infuriated by its depiction of them as inherently undesirable, the butt of an extended arranged-marriage joke.

Perhaps all of that sounds paranoid in the era of K-pop idols and hunky screen stars such as Harry Golding, Ross Butler, and Daniel Dae Kim, but in an analysis of its user data, the dating site OkCupid found that Asian men were consistently rated as least attractive by women, even women who claimed to be open to partners of any race. The pattern grew stronger rather than weaker over the five years the site tracked it. As I wrote earlier, hookup apps, too, both queer and straight, are rife with sexual racism, often masked as "preference." One Asian guy I talked to recalled matching with a woman who told him, "We could be friends if you want, but, no offense, I don't date Asian guys." "How is that '*no offense*'?" he asked. Another guy was asked by a girl where he was born; he told her a small town in Illinois. "She wrote back, 'Hmmm. But where are you *from*?' I'm from Illinois, I'm American, but she kept asking. So I finally said, 'Are you trying to find out my ethnicity?' Next message I got was from the app administrators saying she'd been removed for inappropriate behavior. I still don't know if that was why." Those gender gaps play out in intermarriage rates as well: over a third of Asian American women outmarry as opposed to about a fifth of Asian American men; among African Americans, again, the reverse is true, with black men marrying people of other races at fully twice the rate of black women. Dating cross-racially, then—and specifically dating white women—can have different social and political meaning for a guy like Xavier than it would for one like Spencer, as can dating women of their own races.

College campuses are not immune from those biases. Y. Joel Wong, a professor of counseling at Indiana University, found that both male and female students routinely ranked Asian American men as less physically attractive, less athletic, less aggressive, and less sexual than others. In interviewing a racially diverse group of female students about hookup culture, Nicole Chen, a senior at the University of Michigan, found that while the women held no particular sexual stereotypes for white men, they believed (whether or not they had personal experience) that African American men were more sexually dominant with larger penises than others and Asian Americans—whether of Southeast Asian, East Asian, or South Asian descent—had "small dicks" and were "worse in bed." Putting aside, for a moment, whether one's appeal as a casual partner ought to be the measure of sexual success, or whether penis size has any relationship to women's sexual satisfaction, it's clear that African American and Asian American men are flip sides of the same racialized, gendered coin, with white men controlling the toss (heads you lose, tails I win): one group made hypersexual, the other hyposexual; one threatening for its supposed intellectual inferiority, the other for its alleged superiority. As Wong told me, "Masculine norms don't exist in a vacuum. They reflect the dominant culture, and that is who they are designed to benefit. Men of color are treated based on stereotypes relating to white, heterosexual masculinity: the term to describe it is 'gendered racism.'"

Latinx guys, too, talked about the impact of diverging or conforming to dominant ideals. Mauricio, eighteen, whose parents were immigrants from Ecuador, told me he felt perpetually "disconnected" at the largely white high school he attended. "I don't fit the mold here," he said. "I'm not white, tall, posh, or fit. So I've never hooked up with anyone." The problem, he added, was not "my last name per se," but his comparative height, weight, and skin color. "There's a guy on campus who's Hispanic but if you met him on the street you'd think he was white," Mauricio said. "And he's very much engaged in hookup culture. Latino men who look like me? It's much harder. There have definitely been times where you just feel unwanted, and you start to think, *What if the whole world is like this? What if no one will ever like me?* It feels really . . . lonely at times."

Racial slurs have their roots in the white mainstream—the notion of Asian men as less masculine initially took hold during the Yellow Peril panic of the nineteenth century, based on immigrants' hairstyles and traditional tunics, and was used to justify both the Chinese Exclusion Act and murder—but minority groups can perpetuate them as well. In 2017, on his eponymous talk show, Steve Harvey, who is black, guffawed over a book titled *How to Date a White Woman: A Practical Guide for Asian Men*, speculating it would contain only one page that read, "'Excuse me, do you like Asian men?' 'No.' 'Thank you.'" He then imagined an equally slim sequel, *How to Date a Black Woman*, which would consist of "I don't even like Chinese food, boy. It don't stay with you no time. I don't eat what I can't pronounce." Eddie Huang, a chef and television host, shot back in a *New York Times* opinion piece: "[Every] Asian-American man knows what the dominant culture has to say about us. We count good, we bow well, we are technologically proficient, we're naturally subordinate, our male anatomy is the size of a thumb drive. . . . [There] were times I thoroughly believed that no one wanted anything to do with me. I told myself it was all a lie, but the structural emasculation of Asian men in all forms of media became a self-fulfilling prophecy that produced an actual abhorrence to Asian men in the real world." And when performers turn those racial tropes on themselves? Comedian Ken Jeong, known for his desexualized roles in *The Hangover* and *Crazy Rich Asians*, peppered his 2019 Netflix special with micro-dick jokes and puns on his wife's last name (which is Ho). Whether he is (as he insists) a fierce satirist or a yellowface minstrel depends, in the end—much like Lil Dicky—more on the impact of his performances than his intent, and that is where the problem lies. I'm not so sure the thirteen-year-olds cracking up at *The Hangover Part II* get the metatext when Jeong's character's penis is mistaken for a shiitake mushroom.

Sometimes, Spencer mused, he found himself less angry about the jibes at Asian masculinity than at the "twenty-five out of thirty" of his peers on his college campus who seemed to embody them. "You almost become racist to your own kind," he said, "and that's something I can't help. It just disgusts me, but that's how I feel." In middle and high school, he tried to defy stereotypes by playing multiple sports (even though he felt "automatically at a disadvantage as a short person") and dating white girls.

"I definitely ignored a lot of Asian people," he said. "I was proud of myself for being different, for standing out." Establishing your exceptionalism, though, only goes so far. There was the time recently when he and another Asian guy got separated from their non-Asian buddies at a concert. When they pushed forward, trying to rejoin them, a couple of white dudes responded angrily until they realized what was happening. "Oh," one said, "you're with *them*—we thought it was just you two strange little fellows."

"That really, really pissed me off," Spencer recalled. "It was like, 'Are you saying that because I'm short and I'm really high right now so maybe I seem strange? Or are you saying that because, you know, I'm *Asian* and so you think I'm strange?'" Spencer can never know for sure—that's the thing. Maybe someone's response to you is racist or gendered, or maybe it's just about you. But, as for Xavier, it's the constant wondering, the energy expended on uncertainty, that is ultimately corrosive.

At Any Moment Your Life Can Be Taken Away

The first incident happened just before Xavier's freshman year of high school. I could corroborate some, though not all, of the details, but the specifics are, perhaps, less important than the aftermath, the impact it had on Xavier and his black male classmates. A senior guy, who was also black, got drunk at a party and groped a white girl. The next day, he texted her to apologize, or maybe he wrote her a letter, and somehow that admission of guilt made its way onto social media, incriminating rather than exonerating him. The school administration found out—some say she was reluctant to report him, but her parents insisted—and after an investigation, the boy was expelled. The second case occurred at the end of the following summer, at another party. A white girl had fallen asleep in a bedroom, allegedly waking up to find a boy of color on top of her attempting to remove her clothes. Several months later, she reported the incident to the school's administration, and, again, after an investigation, that boy, too, was expelled. Later, at a school-wide meeting in which students were invited to speak on any topic, the girl read a prepared statement about the assault, claiming the only reason the perpetrator wasn't in jail was because he was black. Xavier was stunned by her insensitivity, plus, he said, no adult intervened. Some of the students of color, including him, walked out of the auditorium in tears.

"It was a shit show," Xavier recalled. "I don't think I've ever cried so hard in a public area. I felt attacked—this was supposed to be a place where I feel comfortable and nurtured, and I was just being torn down to my basic identity. After she was done, it was hard for me to look at my peers and teachers and the head of school, who were supposed to be there for me. I was like, 'In this moment, you all failed me.'"

Xavier didn't necessarily dispute that the two boys engaged in misconduct. His question is whether they would still be at the school—whether they would have been given lighter punishments and second chances—if they had been white and wealthy with powerful parents who paid full tuition (or if the girls they harmed had *not* been all of those things). He doesn't know the answer, nor do I—though, he pointed out, no white boy was expelled for sexual misconduct during his seven years there. On a college level, while no national statistics break down accused perpetrators by race, there's some evidence that black men at predominantly white institutions—as well as foreign students from Africa and Asia—are disproportionately reported for sexual assault and more likely than others to have those charges referred to formal hearings. (Meanwhile, women of color are both more likely to be assaulted than white women and less likely to report it; they are also less likely to be believed when they do.) It's hard to know fully what to make of those statistics, and they are easily distorted by those who want to roll back progress in addressing campus rape, people who may be less concerned with reducing implicit bias than upholding the status quo. It may be that white girls are more willing to recognize and name misconduct by boys of color, and feel more confident that they will be believed if they report it. For Xavier, though, the imbalance triggers the specter of the black brute who rapes "pure" white women. History is, after all, replete with black men who were publicly lynched by white mobs for supposedly sullying white women's virtue. The torture and murder of fourteen-year-old Emmett Till, accused (as it happens, falsely) of whistling at a white woman, helped spark the civil rights movement of the 1960s. As recently as 2015, when white nationalist Dylann Roof massacred nine African American churchgoers in South Carolina, he reportedly said, "You rape our women, and you're taking over our country, and you have to go." In 2018, white supremacist speakers at a Washington, DC, rally referred to the "interracial rape of white women" as one of a litany of alleged civil rights violations against Caucasians.

Rape allegations have been used as a means of social control of African Americans for generations, but rape itself is a tool for social control of all women everywhere. That can put feminists, particularly white feminists fighting to have assault charges taken seriously, at odds with black men; it also puts black women in an untenable position, wanting to protect men of their race from further trauma, even, at times, at their own expense. In a 2019 nationally representative survey, 35 percent of black women said they'd been sexually harassed or assaulted in the past six months, the majority by someone they knew. Even seemingly clear-cut cases of abuse by black men reflect those community tensions. After Bill Cosby's arrest in 2015 for allegations of assault going back decades, a number of high-profile black artists rushed to his defense—Whoopi Goldberg, Phylicia Rashad, MC Lyte, Azealia Banks, Dave Chappelle—some suggesting the charges were an orchestrated attempt to destroy the legacy of a man of color who'd risen too high. And in the final episode of Lifetime TV's six-part documentary *Surviving R. Kelly,* Chance the Rapper admitted, "I didn't value the accusers' stories because they were black women"; just before the show aired, he tweeted, "Any of us who ever ignored the R. Kelly stories, or ever believed he was being set up/attacked by the system (as black men often are) were doing so at the detriment of black women and girls."

The part bias played in the expulsions at Xavier's school may be unknowable, but the suspicion itself was destabilizing. As a young man of color, he already carried the fear of losing his opportunities, even his life, for walking, talking, shopping, reaching for his ID, or simply existing while black. Now he felt unsafe among his white classmates as well. Every guy of color I interviewed at his high school expressed similar sentiments; those who did go to parties, like Aidan, the Lauryn Hill fan, arrived with an entirely different set of concerns from the white students who'd sat next to them for years in class. "It feels like you always have to be on edge and watch yourself," he said, "because you never know who you can trust and at any moment your life could be taken away. And then, when you have this added pressure of only seeing males of color being expelled . . . It can really mess with you. It just really leaves you in this place of anxiety and questioning every decision. I have a pretty strong moral standard, and I have also had it pounded into me—this is consent, this is what it looks like. But still. You never know."

Like Xavier, Aidan dropped out of the party scene, partly because of its reliance on alcohol. "As a black man, it feels like a threat to my life in the most basic way to be intoxicated if I hook up. That has been drilled into my head by my mother and by what I saw in high school.

"I don't want to make it sound like I don't take the same precautions or care as much when I'm with African American girls," he added. "That's not it. But in that case there's a common understanding of where I'm coming from when I say, 'Is this okay? Is this okay?' With white girls, they get impatient. They're like, 'Go ahead. Just *do it* already. Stop asking!' And that raises my anxiety. Because you can't begin to understand what happens if I just 'go ahead.'"

Coloring the Place Up

"To this day, I'm still very thankful to my high school, though," Xavier told me. "It would be very ungrateful and overprivileged if I wasn't. They definitely prepared me for college, academically and even social justice–wise. I think I still would have rather had that than any alternative. My time in high school was definitely . . ." Xavier paused and smiled warmly. "It was *nice*."

Xavier and I were standing in his tiny college dorm room: he and his roommate had lofted their beds and pushed their desks, a TV, and storage boxes underneath, along with the requisite square mini-fridge, a hot plate, and a veritable pantry of prepackaged food, including the largest jar of bouillon cubes I'd ever seen. Xavier was dressed in a gray sweatshirt and black jeans with an orthopedic boot on one foot; he'd sprained his ankle six weeks earlier during a club soccer match, the morning after he broke up with his high school girlfriend. She'd been visiting for the weekend when he mislaid his phone in his room and asked to borrow hers to ring it. It turned out that she'd changed her password and, for the first time since they'd been dating, hadn't shared the new one with him. That got him thinking about the posts he'd seen on Instagram, the ones that he'd wanted to ignore, in which she was cuddled up "a little too close" with another guy, or the time he called her at one in the morning to find her walking home from a party with someone else. So later that day, although he knew it was wrong, he went through her texts while she was taking a nap. Seek and you shall find: in multiple messages to her girlfriends, she described her sexual exploits with other dudes. Xavier didn't react well. He exploded, yelling at her, demanding to know who the men were, kicking his futon couch so hard he broke its hinge. "I lost my temper more than I would have liked to," he admitted. "I wish I had been more calm and sat down and talked it through a little better. I think we probably could be in a different place now—but I don't know if it would be a *better* place. It was a part of me she'd never seen before. So that kind of scared her."

Within a few hours, she was on a flight home and Xavier had drunk himself "into oblivion." He vomited a couple of times, stumbled around his dorm telling anyone who would listen how his girlfriend had betrayed him—although that would be the only time he'd confide in anyone at school. He fell into bed at four a.m., then got up a few hours later for the soccer match, where he was injured. Now every time he looks at his foot he remembers that night. For a few weeks, he said, he was too depressed to do anything: he isolated himself and his grades dropped. "It's gotten better lately, though. My mom's concern was like, 'I don't want you becoming a man-whore now.' Which I could definitely see. Because you're looking for something to kind of fill that void.

"But I don't think I'd be interested in the hookup culture here. And I don't think girls would be interested in me. They're looking for, I don't know, like maybe a John Travolta?" That took me aback; a John Travolta reference felt dated by about forty years, as out-of-touch as the image his white classmates probably had of him. Xavier smiled when I pointed that out. "You know what I mean, though," he continued. "Like that average white male. I think the idea of hooking up with a black male for them—he'd have to be, like, a demon athlete, or something in that realm. Because that fetishization is definitely there. So you have to be on a whole other level to be seen as even average. Or else they're into that kind of risqué idea of being with someone who isn't of European descent. That's not what I want."

What about African American girls? I asked. "There aren't very many here," Xavier said, "and the ones who I know are some of my closest friends. That's also not something I want to do right now. So I haven't hooked up with anybody."

We headed out to meet Emmett at a Chinese buffet, the kind of student dive known for the quantity, not the quality, of its food. On the way, Xavier mentioned that he had liked the idea of attending a historically black college. When he visited Morehouse as a senior in high school, "It felt very empowering. The idea of learning from other black people who know better how to teach me and kind of mold me into the person I need to be—that would have been amazing." But he was offered a full ride at this school, where the African American student population was under 5 percent, and the prospect of graduating debt-free overrode the hesitation about its dearth of others like him, especially since he still had the cost of medical school ahead. Emmett, who was also here on scholarship, was less sure about his future path. He had intended to major in finance but felt conspicuous and lost at the business school, where he never saw another black student. "I have to kind of blame myself for that, though," he said. "I knew what I was getting into. Certainly, I can handle it, but I do feel very out of place. Which is probably not a good mind-set. I have definitely had a bunch of thoughts about quitting, and if my parents had not spent so much time and money on me, like, if I didn't have a bunch of family members that cared, I would not still be here, to be honest."

So here they have stayed, and it's been okay, although neither of them love it. Their classmates hadn't said anything that the two consider blatantly racist (with the exception of a guy who "apologized" to Emmett at a frat party for not having "put watermelon in the punch"), but there had been a steady stream of little things. "Like there is this one girl in my dorm, she always says, 'I'm not trying to be racist or anything, but . . .' And I'm just like, 'Why do you want to preface it that way if you don't feel as if it's a problem in the first place?' Or there's the idea of 'I want to play in your hair.'" Xavier scooted back his chair and waved his hands, as if trying to escape. "You're not going to go play in Timmy's hair. So why are you playing in my hair? I should have that same personal space as anyone else."

Both guys emphasized, repeatedly, that they didn't think of their classmates as venal, just ignorant, thoughtless. Still, the comments stung. They also piled up: small slights—microaggressions—have been found to contribute to conditions among African American college students as wide-ranging as anxiety, stress, depression, suicidal thoughts, hair loss, diabetes, and heart disease, as well as to their high dropout rates. And it turns out that black male students who hew to conventional masculinity—the equation of manhood with suppressing emotion—are even more vulnerable to the negative impact of that everyday racism.

Emmett, for instance, who was under six feet tall and scrawny, was constantly mistaken for an athlete. "At least Xavier is built," he said. "Look at me! There's no reason for me to be confused for an athlete. But I have kids come up to me and swear on their lives that they have seen me on the basketball court. I'm like, 'Bro, I've never even gone to a game.' But to me, someone thinking I was on the basketball team compared to the guy making that watermelon statement . . . those are two different situations. One is just ignorance, and the other is . . ." He trailed off.

Emmett partied less in college than he had in high school—that, he said, was always his plan—but he wasn't averse to hanging out with white "frat dudes" now and again. He'd even checked out a couple of rush events, but aside from the opportunity to make connections that might lead to career advancement, he didn't think Greek life was for him. "I was probably one of four black kids out of about seven hundred rushes this semester," he said. "It was not a good look. They did strive to make it seem friendly. They try to treat you better because—"

"They want to color the place up," Xavier cut in.

"Yeah," Emmett agreed. "They are really trying to make themselves look better. But, you know, some of them—like, you Google SAE and the first thing that comes up is how they'll never have a black man there." That's true: the University of Oklahoma chapter of Sigma Alpha Epsilon, the country's largest fraternity, made national news in 2015 when its members were caught on video belting out a racist ditty; "black man" is a euphemism for the word they actually used. It turned out they had learned the song on a leadership cruise (the lyrics of another version promote violence against women). In 2016, a brother at Yale's SAE chapter was accused of chanting "white girls only" while screening guests for a party and only admitting blond women (racism and misogyny tend to coexist: over just a few weeks in 2014, other campuses' members of that fraternity, colloquially known as "Sexual Assault Expected," were named in rape allegations at Emory, Iowa State, and Johns Hopkins, and at an off-campus house at Loyola Marymount). SAE is not unique. Also in 2016, a notebook belonging to the North Carolina State chapter of Pi Kappa Phi surfaced, containing such lines as "That tree is so perfect for lynching," "If she's hot enough she doesn't need a pulse," and "It will be short and painful, just like when I rape you." During one week in April 2018, two separate frats at Cal Poly were suspended for racist incidents, and at Syracuse University, brothers at Theta Tau were caught on video during a "roast" making racist, homophobic, and anti-Semitic comments as well as simulating the sexual assault of a disabled person. A frat at Xavier and Emmett's university had been suspended two years before, after members shouted racial epithets and threw beer bottles at two black female students walking past their house. At a certain point, one has to wonder whether the issue here is a "few bad apples" or, as sociologist Matthew Hughey, who studies the role of race in fraternities, has said, "more of a rotten orchard."

"I do go to the parties sometimes, though," Emmett said, reiterating that, as a black guy, he was automatically considered the coolest dude in the room. "But I don't really like drinking with those people. Or [doing] anything else with them. Sometimes I'll get really bad anxiety at those parties because . . . You just feel eerie. Anything can go wrong. And if it does, you're screwed.

Because if you're the only black kid at the party, then you're the only black kid at the party, you know what I'm saying? So if the party is shut down by cops, they'll be asking, 'Why are *you* here?' and they won't think that you actually go to this school. Or what if I accidentally break something? When you're participating in extracurricular activities with people you don't trust, it just doesn't make for a really good time."

Emmett has thought about joining one of the handful of historically black frats at the university but isn't sure the time he'd need to put in as a pledge or what he described as "a more intense hazing process" would be worth the effort. None of those frats have houses on the row, either, which is fairly typical for black Greek institutions—they tend to be underfunded—nor do they have the equivalent hookup scene. African American students are, in general, less likely to drink to excess than whites and more likely to endorse gender equality, negating two essential preconditions for casual campus sex. Additionally, while members of both black and white fraternities objectify female classmates, a comparative study found that they diverged when it came to values regarding dating and intimacy: black Greeks were more likely to say companionship, sharing, and love were the main benefits of a serious relationship; white Greeks cited regular sex and being able to skip the condom. The black men also believed that "respecting women on campus" meant "[treating] a woman the way I would want to be treated" (though some health educators would say that is still the wrong standard—you should treat someone the way *they* want to be treated); the white guys said such things as "We won't take advantage of them if they're wasted," or "I'll never ask if she needs a ride home after we hook up. I'll let her bring it up or let her spend the night."

The contrast was not solely a function of race; the researchers believed their findings might have been different had they also interviewed men at a historically black institution. They concluded that the small size, insularity, and visibility of the black community on a predominantly white campus had contributed, at least in part, to the sense of personal accountability that was largely absent for white men: black Greeks were more conscious of their position as leaders and role models, and that extended to better treatment of women.

Emmett had hooked up with a few girls this semester, again almost all of them white. "But I've stayed in pretty safe situations," he said. "The only real out-of-the-blue approaches I've had at parties were drunk white women who I guess are extremely horny for some reason, and I'm like, 'No, this is not okay. Maybe you can hook up with a white guy, but if you're intoxicated, I'm not going to.'"

Sometimes, Emmett said, he suspected he was a bucket list item for those girls: sex with a black guy, another way of experimenting in college, a story to tell their friends. "I've never really taken offense to it, though," he said. "I mean . . . There's not much I can do about it. I don't know how you can change that kind of mind-set." I'd heard something similar from other boys of color I'd interviewed. A freshman at an East Coast college went to the dorm of a white girl he barely knew after she Snapchatted him for a booty call. "It was . . . nasty," he said. "When someone sees you as an object it's just limited what you can do. I felt this pressure to live up to this image of the black male, that sexual prowess and hypersexuality that you see in media and in porn." Another guy told me that he felt overwhelmed by the sexual prospects in college. "As a black man at a predominantly white institution, there is so much sex coming at you," he said.

"For one thing, there are so many more black women than black men, and they know that. Not in a bad way, but it's like a game of musical chairs, and they don't want to be left out when it stops. And then the white women are after you, too. It was too much for me; I met my girlfriend within forty-eight hours of freshman orientation and stayed with her for four years."

I asked Emmett how he imagined his intimate relationships might unfold through the rest of college. He shook his head. "I don't know," he said. "Maybe I'll meet someone back home in the summer, when I'm with people that I trust. But, you know, I don't have really high expectations for any of that here. Not the way those white kids do. Because I never thought college was going to be that for me."

A few hundred miles away, Spencer brooded over the reasons for his own change of circumstances, about why he was so much less popular with girls now than he'd been in high school. "Have I really changed that much as a person that people no longer find me desirable?" he asked. "Or is it that back in high school I wasn't that much shorter than everyone else? Or is it because people are just a lot less accepting of Asian men here? I'm not sure. And I seriously question that every day. I start to doubt myself, you know?"

CHAPTER 6

I Know I'm a Good Guy, But . . .

On a foggy summer morning, I met Liam, eighteen, for breakfast at a San Francisco café. He had just graduated high school and was heading to college in North Carolina. He was a slender, studious-looking boy with dark hair and oversize plastic glasses—if it weren't for the trail of hickeys on his neck, a souvenir from the previous night's hookup, I would never have pegged him as a player. Liam described himself as "athletic but not varsity material." That's why, he figured, his surest route to respect once he got to high school was to hook up with as many girls as possible. "There was a hierarchy among guys at my school," he explained, "and it was completely based on sports, looks, who you're friends with, and your sex life—and on those things alone. Personality? Not really. Maybe if you're a funny guy, then you're 'the funny guy,' whatever. So bragging about how many girls I'd hooked up with, joking about it, was definitely a way to gain status. And I did that, I admit it." When we met, his "body count" was fifty-one. "I only had sex with four of them, though," he said. "Mostly it was making out. But I did go further than that with a lot. And that got me so much status in every aspect of my life. Doing it was physically pleasurable, too, of course, but it was more about buying into the general culture. It was, absolutely, a competition with other guys."

The secret to Liam's success, he said, was all about confidence, or at least the appearance of it. "Even in places like the Bay Area, which is super liberal, super feminist, super all about female empowerment, when it gets down to business, there's still the expectation that the guy will make the first move or nothing is going to happen. It's fine some of the time, but I think what's not really talked about is that guys get really nervous, too, and being intimate with someone can be nerve-racking for both parties. So the reality is that confidence is mostly artificial: it's just about telling yourself, 'I'm confident,' and acting on that. It's not necessarily *actual* confidence."

Liam couldn't recall the names of most of the girls he'd been with, couldn't describe what they looked like, the sound of their voices, how they kissed—anything about them, really. All he knew was his number. He'd met most of them at youth group conventions or summer camp—he avoided casual hookups with classmates, because his public high school was relatively small and things could get messy. There'd been nights when he'd hooked up with six different girls, grinding on them, then turning them around on the dance floor to make out, sometimes leading them to a spot where they could progress to more. It was exciting, he said, but not particularly . . . He struggled for a word and landed on "personal." That's how it had been from the start for Liam: he described the first time a girl performed oral sex on him, during an eighth-grade camping trip, as "traumatic" because everyone quickly found out and ribbed him about it. "It was genuinely hard to deal with everyone knowing such intimate details about my sex life," he said. "And it didn't give me status. Not then. Because I didn't play it that way." Better, he decided, to turn it around and use the rumor mill to your advantage. More recently, as high school was ending, Liam was experiencing a change of heart: he'd been trying to develop "relationships with girls as human beings and not numbers, which, again, I'm not proud of that, but then I must have been, because I did *brag* about it."

Although he's treated girls as more or less disposable, Liam talked about being "communicative and compassionate" with his partners—even when he was drunk. He's the guy, he said, who "calls out other friends for being shitty, for being disrespectful to girls." And he always tries to check in verbally with partners, to ask if they are comfortable.

So what, exactly, do you say? I asked.

"Well, something like 'Do you want to go down on me?' "—a question that, as opposed to, say, "How do you feel about oral?" some health educators consider more of a directive than a true conversation starter. "But sometimes it would be more wordless. Like, not talking about it, just going further."

I asked Liam if that included pushing a girl's head down to get her to give him a blow job. He nodded. "Yeah, I've done that. I have. Not proud of that, but I have."

Liam shifted uncomfortably in his seat. "I think of myself as a good guy," he said. "I *am* a good guy. But . . ." He looked pained, seemed to wrestle with himself, then sighed. "I didn't think I was going to talk about this, but, sure, why not?"

There was this one time after his junior year of high school, he said, when he hooked up with an "absolutely gorgeous" girl at camp. One thing led to another. He went down on her; she went down on him. Then they moved on to intercourse. Or, to be more accurate, Liam did. The girl didn't say no to the act, but she also didn't say yes. Nor did they talk about it afterward. Liam was left uneasy, concerned that she hadn't wanted to do it. They did hook up a few more times, and each time she clearly consented to intercourse, but then she abruptly broke things off without saying why; she just stopped talking to him. At first, Liam felt victimized. "It hurt," he said. "It sucked. And the pressure I felt as a guy was, 'You had sex with her, so you can't complain.' But the truth was, she fucked me up. She made me really sad, confused, whatever. And I didn't recognize that the fact that I took advantage of her could have been the reason why she ended things."

He tried to put the incident out of his mind, and for months he had, more or less—he went home and became busy with school and friends and college applications. But lately, the memory has been troubling him. "I did take advantage of her," he said. "I did. And part of me is like, 'Oh, I shouldn't have done that.' And I regret it. But there's also a part of me that's confused. Because I know people who have really been taken advantage of in a forceful, aggressive manner. This seems different—I think it *is* different. But I don't know . . . I don't know what I'm supposed to feel. I don't know what I'm supposed to do."

"No One Ever Thinks They're a Rapist"

The latest debates over sexual consent have been raging on college campuses since 2011. That's when the Obama administration issued a "Dear Colleague" letter to university officials reminding them they were responsible for upholding all aspects of Title IX, including those that involved sexual misconduct: at the time, three quarters of schools were reporting their assault rate at an improbable zero. In reality, according to the AAU Campus Climate Survey, which consisted of one hundred fifty thousand students at twenty-seven public and private universities, 23.1 percent of female students had been sexually assaulted through physical force, violence, or incapacitation since entering college, including 10.8 percent who experienced penetration. By 2014, the state of California had mandated that all universities receiving public funds use the "affirmative consent" standard in sexual misconduct hearings (in a separate legislation, public school district health curricula were required to include consent education, although there are no guidelines as to what that might entail). Sometimes called "yes means yes," the law requires a sexual partner to obtain "affirmative, conscious, and voluntary" agreement at each stage of an intimate encounter. New York passed similar legislation a year later, by which time eight hundred campuses nationwide were using that standard in adjudicating complaints. That's a dramatic break from the past, when, until and unless a person said no (and sometimes not even then), anything was fair game.

When I first began interviewing high school and college students about sex in 2010—before the "Dear Colleague" letter, before Donald Trump's "locker room banter," before the fall of Harvey Weinstein, the rise of the #MeToo movement, the conviction of Bill Cosby, the accusations against Brett Kavanaugh—most, whether male or female, believed rape was something perpetrated by strangers in dark alleys. So the fact that it even occurred to Liam to be concerned over his actions is strangely heartening: a sign that young men are grappling (if some more sincerely than others) to integrate ideas about gender, sex, and power that may contradict previous, deeply held expectations. That's major progress in a very short time, and a long way from

1993, when the *New York Times* scoffed at Antioch College's pioneering affirmative consent policy as "legislating kisses."

Some of the guys I talked to had found the erotic spark in "yes means yes." Recall Wyatt, the "feminist fuckboy" who said, "I'll just put it out there—affirmative consent is really hot. It's exciting to have a girl saying, '*Yes!* I want you to do this.' '*Yes!* I want you to do that.' To feel she's really into it. It's a pretty awesome thing for both of you to have that sort of connection."

Others were less motivated by sensuality or personal ethics or the fundamentals of human decency than by a desire to avoid getting in trouble. Zachary, a high school senior also from the Bay Area, outlined a philosophy that, at best, seemed to miss the point. "If the girl wants to go further the first time we hook up, I'm like . . ." He lifted his hands in the air as if he'd been burned, then glanced downward, presumably to where a girl's head would be. "She can do whatever she wants. I'm fine with that. But I'm not going to go there because I'm paranoid as fuck."

In practice, though, young men who do understand the new standard and consider it fair don't necessarily follow it, even when they believe they do. In 2015, Nicole Bedera, a doctoral candidate in sociology at the University of Michigan, conducted in-depth interviews with sexually active, heterosexual men attending a college that provided multiple educational programs on affirmative consent. Nearly all of them could offer at least a rudimentary definition of the concept—such as both partners wanting to be doing what they're doing—and most endorsed the requirement of active, conscious, continuous, and freely given agreement by everyone engaging in sexual activity. Over a third further claimed that they regularly established verbal consent. Yet, when asked to describe their most recent experiences in both a hookup and in a relationship, only 13 percent had, in fact, engaged in conversation about their intentions, and in the vast majority of those cases, their female partners had initiated the discussion.

When the men realized that their actions conflicted with their stated beliefs, they expanded their definitions of consent rather than questioning their behavior. Their ideas of "yes" turned out to be so elastic that for some they encompassed actions that met the legal criteria for assault—such as the guy who had coerced his girlfriend into anal sex (she had said, "I don't want to, but I guess I'll let you"). She made it clear that he should stop. "He did, eventually," Bedera told me, "and he seemed aware of how upset she was, but he found a way to rationalize it: he was angry with her for refusing him because he thought a real man shouldn't have had to beg for sex." Perhaps that should not be surprising, given that men who are incarcerated for committing old-school rape—those stranger-in-a-dark-alley guys—also typically believe that their victims wanted it and for many of the same reasons: moaning (which can indicate pain as well as pleasure), accelerated heart rate (a sign of terror as well as excitement), vaginal lubrication (which is involuntary). They, too, generally deny "rape," even if they acknowledge the more euphemistic "nonconsensual sex."

The truth is, research has shown young men to have a remarkably sophisticated and subtle understanding of sexual refusal, regardless of whether a partner ever utters the word "no"; that renders dubious the common defense that they "can't tell" or "aren't mind readers." What's more, where "yes" is concerned, guys seem downright clairvoyant. Bedera's subjects considered such nonsexual physical cues as direct eye contact to be clear propositions. Obviously, looking into someone's eyes doesn't always signal seduction, and guys know that. When

I, a middle-aged woman, would look straight at them while asking about, say, cunnilingus, not a single one mistook that for a come-on. Yet, when my eighteen-year-old female intern conducted interviews at her college, every guy hit on her: the same questions from an attractive peer were sexualized because the boys wanted them to be. Some guys Bedera talked to believed smiling indicated a girl wanted to have sex: it might, though people smile for all sorts of reasons, including discomfort and appeasement. Compliments were another frequently cited indicator (causing me to reconsider expressing admiration for a male passenger's stylish luggage the last time I was on an airplane). Standing close. Dancing. Touching someone's arm during conversation. A third of the supposed sexual signals came up only once in Bedera's study, making it hard to ever predict what a man might see as an invitation: a random emoji; the lack of a bra; sitting on a guy's lap in a crowded car. True, any of those *might* have meant sexual interest—or not. The only thing they all had in common was that the guy in question read them as evidence. The boys also tended to equate enthusiastic participation in any sexual act (such as kissing) with enthusiastic consent to vaginal intercourse.

When they drink, young (and not so young) men are even more likely to overestimate female sexual interest—as well as to overstate women's role as initiators—interpreting *any* expression of friendliness by a girl as: *It's on.* In 2016, researchers at Confi, an online resource dedicated to women's health issues, asked 1,200 college students and recent graduates nationwide what they would "expect to happen next" if they went home with someone whom they'd met and danced with at a party. Forty-five percent of the men considered vaginal intercourse "likely"; only 30 percent of the women did. The figures were similarly skewed for oral sex. Additionally, one in four men said women would need "some convincing" in order for sex to happen; this was not twenty or thirty years ago, mind you: this was 2016. Those perception gaps are a setup not only for assault—out of unexamined entitlement, if not flat-out ill will—but also for boys' subsequent denial of responsibility and, quite possibly, their claims of false accusation. According to the same survey, men found the actions of a "tipsy" guy "much more acceptable" than a sober one, meaning they let themselves off the hook for potential sexual aggression, even as female assault victims who drink are blamed. (Women project that bias on one another, too: in a 2019 survey of college students in the Midwest, both sexes believed a woman holding a drink in a social setting was more "sexually available" than a woman drinking water or than a man drinking alcohol. They were also less likely to say they would help a woman in a risky situation who had been drinking because they'd presume she was interested in casual sex, so the situation was not a threat to her.)

Despite all evidence to the contrary, we—men, women, adults, teens, and, perhaps especially, parents—still want to believe that only "monsters" commit assault. True, it may now be monsters we know—our employers, our clergy, our classmates, our teachers, our favorite celebrities, our politicians, our Supreme Court Justices—but they are monsters, nonetheless. A "good guy" can't possibly have engaged in sexual misconduct, regardless of the mental gymnastics he has to go through to convince himself of that. Even men who admit to keeping sex slaves in conflict zones will claim they did not commit rape—it's that other guy, that "monster" over there, that "bad guy," who did. In reality, one of the few traits rapists have been found to reliably share is that they don't believe they are the problem.

Among the young men I've talked to, some that I liked enormously—friendly, thoughtful, bright, engaging guys—have "sort of" raped girls; pushed girls' heads down to get oral sex; taken a Snapchat video of a prom date giving them a blow job without her knowledge and sent it to the baseball team; shot an iPhone video of a girlfriend performing a blow job (again, without her immediate knowledge) for "personal use." They all described themselves as "good guys." And they were, most of the time. But the truth is, a really good guy can do a really bad thing. The cost of admitting that, however, can feel perilously high. Liam had not told any of his friends about what happened at camp because "if you tell someone about something like that, and it's the 'wrong person,' that's a terrifying prospect. You could be labeled as a sexual predator, a rapist, and that can destroy your entire life." So to avoid being labeled a "monster," he felt forced underground, unable to take the responsibility he might want to for his behavior.

Other young men, especially those who are elite and white, use what sociologists C.J. Pascoe and Jocelyn Hollander have called the "'good guy' defense" as a shield. When Brock Turner, a former Stanford swimmer, was convicted in 2016 on multiple charges of sexually assaulting an unconscious woman at a fraternity party, his father famously described the six-month jail term as "a steep price to pay for twenty minutes of action." A letter submitted to the court by one of Turner's female friends, which the judge had considered in his sentencing, also claimed Turner wasn't a "real rapist" because what he did was "completely different from a woman getting kidnapped and raped as she is walking to her car." Turner himself testified that, although he fled the scene after being spotted next to a dumpster violating an incapacitated girl, "[In] no way was I trying to rape anyone." The encounter, he insisted, was consensual (tell that to the victim, or to the two Swedish men who were distressed to the point of tears when they called the police to report the crime). He is a good guy, and good guys don't rape, therefore, ipso facto, he could not be a rapist.

Men learn too often, subtly or overtly, to prioritize their pleasure over women's feelings. That may or may not lead to assault, but it does raise ethical questions over how men treat sexual partners, particularly in encounters that skirt the edges of consent. That's why the accusations against comedian Aziz Ansari in January 2018 caused such a stir. "Grace," a pseudonymous twenty-two-year-old, claimed that when she went back to Ansari's apartment after a dinner date, he tried multiple times to initiate sex with her (mostly by repeating, "Where do you want me to fuck you?"). She declined, both physically and verbally, but he persisted. When she explained, "I don't want to feel forced because then I'll hate you, and I'd rather not hate you," he backed off, inviting her to "just chill over here on the couch." A few minutes later, though, he began playing with her hair, then spun her around and pointed to his crotch, indicating that she should perform oral sex on him. And she did. The encounter continued, him pressing, her reluctantly (and to some, inexplicably) acquiescing; him oblivious to her discomfort, her not wanting to offend. Nothing that happened between them was illegal. Ansari is not a Weinstein or a Cosby. He is not even a Louis C.K., who made female associates watch him masturbate. He was just another overeager guy trying to talk a woman into sex, viewing her limits as a challenge he needed to overcome in order to score. What he did was not unusual and was not, in truth, newsworthy; yet that was the very reason it was news. Because it extended the conversation *beyond* legality, revealing the most banal and pervasive of power dynamics: that men interpret women's behavior through the filter of their own wishes. Their claims of "miscommunication," then, Nicole Bedera concluded in her research on college students, may actually be part of "an expectation that they control both partners' narratives" about a sexual experience, including about consent.

The narcissism of male desire is instilled early, reinforced by media, peers, and parental silence, and by girls who have themselves been trained from an early age to take men's needs and desires more seriously than their own. Brandon, a junior at a Pennsylvania college, recalled other boys chanting his name when he walked into his high school's gym the day after he lost his virginity. "It made me kind of sick to my stomach," he said. "I think a lot of guys felt caught between being externally 'celebrated' and internally confused. If you wanted to party, you had to become numb to it." These days, Brandon has a "good guy" rep with his female peers. They also see him as a safe option for fully consensual, casual hookups, so he's had a lot of partners. But that hasn't necessarily made him happy. "What have I gotten away with that another guy wouldn't have been able to because I'm a 'good guy'?" he wondered. "Like, if a guy orgasms first in a hookup, he's very rarely going to make sure the girl also does. I've done that. I've never heard of a girl pressuring a guy that way. It's like the standard is so low, as long as you are doing the bare minimum of being a decent human being, you are a 'good guy.' And even though I still fall into hooking up drunk or buzzed and not thinking much about consent, I'm still a 'good guy' because what other guys are doing is *so* bad.

"But everyone *thinks* they're a good guy," he added. "They've all been told that by someone—by family or friends. So no one thinks they're a rapist except maybe in the worst situations. And you know what? Most of the time, not even then."

Some of the boys I met were aware that they'd crossed lines, but, especially when their actions fell short of forcible rape, they weren't sure how to address it. Reza, the sophomore at a Boston college I spoke with in chapter two, counted nine high school hookups in which he'd found himself in what he described as "a gray area." One time, he recalled, a girl at a party told him she'd taken a prescription drug that made her woozy, but he chose to ignore that. "It was just in one ear and out the other. I didn't think about it at all," he said. He ended up touching her breast under a blanket. "She was like, 'Yes! Yes!' But when she got up, she couldn't stand straight. And I was like, 'Shit.'" When he texted her the next day to apologize, she didn't remember the encounter and told him not to tell her what had happened. "I felt awful," he said. "I still do. But you're trying to work it out: How aggressive is too aggressive? How much is too much? What counts as okay? My parents always told [me], 'Respect women!' But that's kind of like telling someone who's learning to drive not to run over any little old ladies and then handing him the car keys. Well, of course, you think you're not going to run over an old lady. But you still don't know how to drive."

Sometimes, listening to guys like Reza, I found myself privately dismissing their transgressions as no big deal. Because in some ways they weren't: they were the classic teenage fumbling of someone trying to learn the rules. But what if that learning curve comes at girls' expense? Maybe, as a woman myself, my standards were warped by my own inevitable experiences of violation. I'm not talking about rape, but the years of catcalling or being groped on the subway or swatting men's hands away at a party or staying "safe" on a date. Those were not "microaggressions": they were *actual* aggression, the kind to which one becomes more or less inured. Looking back, I feel lucky that I didn't endure worse. So perhaps that leads me to reflexively dismiss or excuse misconduct up to a certain threshold, though I couldn't tell you what that threshold is or how it compares to someone else's. I still wouldn't want Reza punished, at least based on his accounts of what happened, but I'm glad he's reflecting on his behavior and, hopefully, growing as a result.

Then there was Nick, a Bay Area native attending college in Washington state. Nick was nineteen, freckle-faced and sandy-haired with one pierced ear and a small compass tattooed on his wrist pointing true north. He first learned how to approach girls at middle school dances: "They were things of legend," he recalled, "like a club scene. A giant, dark room with music blasting. Everyone made out there for the first time. In sixth grade, you'd walk up to a girl and tap her on the shoulder and be like, 'Do you want to dance?' But then in seventh grade, there was this unspoken shift where guys stopped asking. You'd just go up behind a girl and start grinding on her. So every five minutes, these girls would have someone's hard dick rubbing up against them out of nowhere. It would not be atypical to see boys trying to shove their hands down girls' pants, because the girls were wearing spandex that cut off at their belly button. And so many girls didn't know how to say no in those situations. They also don't know what's normal and what's not. We were constructing the rules. There was no talking; there couldn't be because it was too loud. It was just totally impersonal, like training for what would happen in high school: the boys would get blackout drunk or cross-faded and just keep pushing the boundaries." By college, he was unsurprised when a fraternity on his campus was suspended for secretly taking photos of women during sex and posting them to a group chat. Guys believed they could just do stuff like that.

Even though Nick resisted those messages himself—he was a "good guy"—they still got into his head. The spring of his sophomore year, shortly before we spoke, he picked up a girl in a bar. They kissed for a bit and exchanged numbers. When they met up again the following night, she invited him back to her place. "To a guy, that registers as, 'Oh, shit, this is happening,'" he said. "Like, 'I've got to focus.' There's so much around performance anxiety. Especially for a guy like me who hadn't had sex with multiple women before college. I want to do the right thing, but I don't know what the right thing is. I just know what I know, which is a lot of really confusing and wrong shit." The "shit" he'd been learning since middle school. "So immediately it was boom, boom, boom, boom, and we got to a point where we were about to have sex, and she put her hand on my chest and said, 'Whoa! I don't want to do that.' And I thought, *What?* I was doing everything I'd been told to do! But in that moment, I could see just how wrong it was. The utter lack of communication that took place in those five to ten minutes. And even realizing that I didn't feel great myself about what we were doing. I just thought that was the only option."

Theo, a sophomore at an East Coast college, described a similar experience, though in his case, he ignored his partner's hesitation: he asked if she wanted to have intercourse, and she demurred, suggesting they "do it Tuesday instead." He countered that he was busy that night and kept going. She remained passive as they proceeded; when he asked if she was okay, she responded first with "I'm not sure," followed by "Um, I guess so?" "I suppose there was something in the back of my head that I wasn't fully listening to," he admitted, looking pained. "I guess when you've been flirting with someone the whole evening and you feel close to what you've been wanting to happen, it's difficult to put on the brakes. And . . . I don't know. I was enjoying myself. I was having what in the moment was a positive sexual experience. I think . . . I think the truth is, I just *wanted* to. Which is scary."

"I Should Call This Girl and Apologize"

Liam never did tell anyone about what happened at camp, including—and maybe especially—his parents. His dad is an architect; his mom a social worker. Liam described them as politically progressive, emotive, and open, and his relationship to them as close. They were tolerant of his drinking in high school (as long as he didn't drive) but were emphatic that he was too young for sex. They were, then, oblivious to his reputation, unaware that he'd been engaging in oral sex since age thirteen and vaginal intercourse since fifteen. Frankly, Liam felt they had done him a disservice. "I love my parents," he said. "They have taught me a lot of things. But when it comes to sex, they haven't. Just about nothing. They haven't guided me, and there've been times where I really wish they had, that they'd given me some advice. I wish that they had told me that sometimes it doesn't work, sometimes it's really scary. . . . Honestly, I just wish they had told me *anything*, because I was sort of thrown into this place where I knew literally nothing except [from] a couple of classes in school and watching porn. And I don't know. I guess I resent them a little bit for that. . . . I mean, it's uncomfortable to talk to your parents about sex, but it's also one of those things that I wish they had forced me to do, because I feel like I would have been better prepared. Maybe I could have not gone into some more uncomfortable situations if they had talked to me."

Despite their apparent mortification, boys do want their parents to talk to them about physical intimacy, for someone to go beyond the classic don'ts: don't have sex, don't get anyone pregnant, don't get a disease, don't be disrespectful. They are particularly eager to have their fathers talk to them about their own experience with sex, love, even regret. But according to a 2017 national survey of three thousand high school students and young adults by the Making Caring Common Project, the large majority of boys had never had a basic conversation with their parents about how to be sure in advance that your partner wants to be—and is comfortable—having sex with you or about the importance of "being a caring and respectful sexual partner." More than 60 percent had never heard from their parents about the importance of not having sex with "someone who is too intoxicated or impaired to make a decision about sex." Neither parents nor teachers of most of the male students had ever told them not to catcall girls or use degrading comments such as "bitches" or "hos," even though 87 percent of the girls reported having been sexually harassed. Those ideas might seem self-evident to an adult, beyond the need for comment, but given the rates of coercion, harassment, and assault, boys are clearly not learning sexual ethics merely by osmosis. What's more, most of those who did have such conversations with adults described them as at least somewhat influential.

Richard Weissbourd, the lead author of the survey, believes that parents have abdicated responsibility for talking with their children about sexual ethics or emotional intimacy, especially their sons. The conversation is more crucial than ever, the report concluded, because we appear to live in a time of "pervasive misogyny and sexual harassment." "If you ask many parents whether it's really important that your son has a lot of integrity and is a good person, they would absolutely say yes. But if you were to ask, 'Have you talked to your son in a concrete way about the many ways he can degrade women?' Most parents, I think, would say no."

As for Liam, he'll never know whether or not he harmed his partner. As someone who has interviewed many girls, I can imagine a scenario in which she did feel violated—women often freeze in response to unwanted advances—and then, to regain some sense of control, had sex with him consensually several more times until she felt in a position of strength before dumping him. I can also imagine that, as a girl socialized to value male needs over her own, she passively consented, even though she neither wanted nor desired sex. It's possible she was fine with it, that she wanted to have sex with Liam as some form of validation or just to see what it was like, and once she got that, she simply moved on. Or maybe she decided she didn't like him or the chemistry was wrong. In so many of the encounters boys described to me, I couldn't know, though sometimes it seemed like the shadow of a girl hovered behind them, a girl who was furious or traumatized or rolling her eyes, one who would have told the same story very differently. The question was how to get the boys to see that, too. "Honestly," Liam said, "I don't know what I'm supposed to feel about it. I don't know if I should call the girl this morning as soon as we finish talking and apologize. That's what I probably should do. But you and I both know I won't."

CHAPTER 7

All Guys Want It. Don't They?

Maybe if Dylan hadn't been so distraught, he wouldn't have drunk so much. Earlier in the day, he'd visited a friend in the hospital who'd been in a car accident, and he was a little freaked out by the other boy's pain. So when he got to the party, a typical Friday-night banger, he downed nine, maybe ten shots of vodka. Then again, maybe he would've done that anyway. He was seventeen, a junior in high school, a straight-A student, captain of the soccer team. He ran with a good-looking, athletic, popular crowd, the kind of kids who defined success by acceptance to the narrowest slice of selective colleges. They worked hard. So when the weekend came, who could blame them for wanting to blow off a little steam: drinking, smoking weed, hooking up?

Dylan dropped onto a couch and immediately passed out. That's where Julia found him. The two had become friends earlier that fall; she was funny and friendly, and Dylan considered them to be pretty close. She shook him awake, at least long enough to drag him, stumbling, to the bathroom.

After that, Dylan's memory fractures. He recalls Julia, who

was sober, fumbling with his pants. Julia, touching his penis. He must've gotten an erection, though he doesn't remember that. His head ached, that he can say for sure. And then . . . nothing. "I had to call her the next day and—oh God—ask if we'd had sex," he told me.

She said they had.

"I didn't want to do that!" Dylan responded. He'd never had intercourse before. He'd wanted the first time to be special.

"Oh, *please*," she shot back. "Don't give me that shit. All guys want it."

Can Girls Assault Boys?

When I began interviewing boys, I assumed, at least for those who were heterosexual, that our conversations about consent would flow in one direction: exploring their own understanding (or its lack) of how to ensure a partner has said yes. So I was surprised by how often they brought up their own experiences of unwanted sex—encounters in which girls did not seem to hear or respect "no" or, as with Dylan, took advantage of them when they were too incapacitated to protest.

At first I dismissed boys' accounts of being victimized. They're bigger than girls, after all, stronger. How hard could it be for them to get up and walk away? Why not drink less? Were they *really* describing assault, or just bad sex? Then I realized how I'd react if someone lobbed those same questions at a girl. Was it so different? Maybe my deeper fear was that surfacing boys' stories might distract from #MeToo, dilute its message of systemic, pervasive sexual harassment and violation of women. But perhaps the opposite is true. After all, the notion that all guys are sexually insatiable—ever ready, incapable of refusal, regret, or injury—reinforces the most retrograde idea of masculinity, the very thing the latest feminist wave is trying to change. Disregarding boys' abuse, whether by other men or by women, risks driving them toward shame and disconnection, and sets them up for potential mental health issues, just as it does for girls. And who among us can judge someone else's trauma, deciding which assaults "count" and which do not? That would be reason enough to question my assumptions, but there is another issue, as well: If guys are supposed to deny their own violation, how can they feel compassion for a girl's? If they can't say no, how are they supposed to hear it?

In middle and high school, boys actually report being the targets of dating violence—including slapping, hitting, and pushing—at rates similar to, and in some surveys even greater than, girls (though when it comes to the most violent acts, such as murder, 90 percent of victims are female). When researchers ask students about nonconsensual sexual penetration, 98 percent of perpetrators are male, but the numbers on other forms of sexual misconduct tell a broader story. A three-year, $2.2 million study of undergraduates at Columbia University called the Sexual Health Initiative to Foster Transformation (SHIFT), concluded in 2017, found that 22 percent of students experienced behaviors that met definitions of sexual assault; 80 percent of them were female, but one in eight male students had also been victimized. Other research has found that fraternity members are disproportionately likely to perpetrate assault; SHIFT revealed they were also more likely to be its targets. "Men, as you might imagine, were more likely to report unwanted sexualized touching," explained Claude Ann Mellins, a psychologist who codirected the study with sociologist Jennifer Hirsch. "But that doesn't mean they're not experiencing penetrative or attempted penetrative assault. And over 50 percent said the method of perpetration involved incapacitation. The other most cited methods were verbal coercion, lying, or threatening. Saying things like, 'I'll tell everyone you're a virgin,' or 'I'll tell everyone you're gay,' or criticizing someone so he'll be more willing to engage. There are always biases going into research, and I was surprised to find the alleged perpetrator in these instances was a woman over 60 percent of the time."

Listening to Mellins, I recalled a college sophomore who'd told me that during his first high school party, at age fourteen, a seventeen-year-old girl led him into a bedroom and performed oral sex on him. He didn't want her to do it but was too drunk, too naïve, and too worried about rumors she might spread to leave. The encounter still confused him. "Like, if it's the guy who didn't consent," he asked me, "what do you call that?"

"Hey, You Got Laid!"

Dylan and I sat on a park bench near his Northern California home, soaking up the late-winter sun. He stretched his legs out on the grass: he was six feet tall and weighed 185 pounds (I know because he told me so three times), with brown hair and frost-blue eyes. Although he was critical of the classic markers of masculinity—he ticked off financial success, athleticism, physical stature, sexual prowess, stoicism—he also aspired to them, which made him feel like a hypocrite. He spent at least two hours a day at the gym, trying to build bulk; sometimes, he said, he worked out until he could barely walk. When I asked what he was trying to achieve, he shrugged. "I don't even know anymore," he said. "I just keep doing it. The ceiling is never necessarily high enough for me, which is kind of a bad thing I guess." Nor was he above bragging to friends about the "chicks" he'd hooked up with, though he drew the line at discussing sex with an actual girlfriend. "Because hooking up is impersonal," he explained. "So you can just be like, 'Oh, dude, she was so hot. I fucked her for, like, an *hour*.'" Aside from their tall tales ("It's always about stamina with guys," he said. "Always."), he believed his friends were pretty skilled at hooking up. At least that's what the girls tell him. "They'll say, 'Oh, he was *so* good,'" Dylan said. "And I take their word for it. I don't know why they would lie about being satisfied."

Telling his friends what had happened to him was unthinkable, at least initially. Nor could he tell his parents. His mother, a devout Catholic, considered sex outside of marriage to be a mortal sin; being shit-faced was hardly an excuse. His father, an atheist, was more lenient, but their relationship was never close. "He's this very stoic guy," Dylan said, "very cold and unemotional, like a 'man's man': the type who works all the time and is quietly dissatisfied. I feel like he's not very happy, and that really hurts." The one thing they do bond over is sports: Dylan's dad was also a star athlete in high school and college; he drilled Dylan on soccer fundamentals every weekend for years, and they still kick the ball around on Saturday mornings.

He could never tell a guy like that that he'd been *assaulted*. By a *girl*.

Instead, Dylan channeled his feelings into death metal playlists on Spotify. He also began, as he put it, "acting like a dick," mocking less popular kids at school and lashing out at his teammates. "If people were being loud in the car I'd be like, 'Be quiet or I'll beat your fucking ass!'" he said. "I got really aggressive.

Honestly, it scared me. I'd get angry, then I'd get sad that I got angry, and then I'd feel like a wimp for being sad about being angry, and then I'd go back to being angry again. It probably would have been healthier to just feel hurt, but it's like I only know two options: happiness and anger. So I chose anger."

No question, Dylan took his experience harder than most boys I'd met, yet he wasn't alone, especially among those who had wanted their "first time" to be special. Leo, a high school senior from New York, told me he'd hooked up with an older girl the previous summer while he was high; she was performing oral sex on him when, suddenly, she straddled him and "put my dick inside her" (recall that physiological responses—lubrication, erection, orgasm—can be involuntary and do not imply consent). He'd never had intercourse before. He tried to move away, but she held on; he told her to stop—twice—but she ignored him. Finally, he shoved her off. Like Dylan, he, too, subsequently became depressed and volatile, subject to panic attacks and verbal aggression. He also considerably upped his use of illicit drugs: weed and street Xanax to numb him out, bootleg Adderall when he had to focus. "I knew it was linked to what had happened," he said, "but I didn't want to admit it to myself."

More commonly, though, boys brushed off their unwanted encounters. A college sophomore who faked "whiskey dick" to avoid sex with a girl who'd talked her way into his bed chalked that up to "a learning experience." Another college sophomore felt slightly "weird" after discovering he'd had intercourse with a stranger—a girl with a reputation for "getting guys really drunk on purpose"—after blacking out at a frat party; his takeaway was that he should drink less. To better understand this phenomenon, sociologist Jessie Ford interviewed forty straight male and forty straight female college students who'd had experiences along a continuum from unwanted sex—defined as sex they chose to continue but believe they could have stopped—to assault; she also interviewed an additional thirty-five LGBTQ+ students. About a quarter of the men described situations in which they'd been too incapacitated to refuse. But most had succumbed to something subtler: a voice in their heads that said, assuming that a girl was neither too drunk nor too "ugly" (both of which were socially acceptable excuses to refuse), guys should always be "down to fuck." Rejecting someone's advances would be awkward, unmanly, "gay." Some feared being rude. "They thought saying no to a blow job would hurt a girl's feelings," Ford said, "or it would make them look stupid or weird. It seemed easier to just go along with it than to end things." That's remarkably similar logic to that of the girls I've interviewed who would go down on guys without wanting to, especially during a hookup. They, too, were exquisitely tuned to gender expectations—the potential to be called a bitch or a prude—and feared seeming impolite if they expressed a direct "no." They, too, would rather feel violated than risk humiliating or disappointing a partner. (Though, notably, girls' physical gratification is not a factor in either scenario.)

It surprised me to learn that, like girls, boys will also fake orgasm to end an encounter, albeit less frequently. In one study of college students, two-thirds of girls reported pretending to climax, but so did a quarter of the boys and for similar reasons: they wanted sex to end; they wanted to make a partner happy; they didn't want to hurt her feelings. For guys I met, faking orgasm was most common when sex was unwanted, perhaps because male release is seen by both sexes as the necessary conclusion for a successful encounter. "I should never be in the position where I have to tell somebody that I'm having sex with that I came when

I didn't," a college junior told me. "But I have a couple of times. It's whack. Once at a party this girl says, 'I'm going with you to your room,' and I know I don't want to hook up with her, but she insists on coming back with me, and she gets into my bed. And I'm like, 'Huh.' I knew I didn't want to, but we start kind of having sex and there's no real interest or chemistry, so . . ."

This is not to say that heterosexual men's and women's experiences with unwanted sex are fully comparable. By their senior year of college, women are still twice as likely as men to have been assaulted and are subject to a wider, more constant range of aggressive behavior. The women Ford interviewed reported incidents that spanned from catcalls to forcible rape. Their accounts also typically included either actual or implicit threats of violence, such as the fear of being killed—men's did not—and it's those feelings of complete helplessness, the sudden turnaround from trusting someone to believing he might murder you, that are most strongly linked to anxiety, depression, and other negative mental health outcomes. Perhaps as a result of all that, fewer men expressed explicit psychological fallout from their unwanted experiences. "They would say something like, 'If this happened to a woman you would call it assault,'" Ford said. "Or, 'Maybe at some point there will be a case of a man coming forward, but I'm not going to be the one to make a big deal about it.'

"Really, though," she continued, "that was all cover for the masculinity issue. What happened to them might be creepy, but because a woman did it to a guy, it's like, 'Hey, you got laid.'"

That's exactly how friends responded to Leo, the boy in New York—plus, *dude*! The girl was nineteen! "I'd laugh about it, too," he said. "No one ever picked up on how much it hurt me. But it really fucked me up. I became afraid of sex. I thought I'd never do it again. The worst was when I was at a party playing Truth or Dare and someone asked, 'Have you had sex before?' And I thought, *Shit! What exactly* is *sex?*"

Among the high school boys who confided in me about unwanted sex, there was often an age difference between themselves and their aggressors. The idea of a mature woman initiating a lad into the wonders of the flesh is a well-worn trope, a staple of both porn and mainstream media: the MILF, the stepmom, the teacher. Mrs. Robinson. Stifler's mom in *American Pie*. The fantasy can blind us to real-life abuse. When news broke of a Florida teacher having sex with an eighth-grade boy, for instance, one social media wit posted, "This poor boy [is] currently in surgery due to the trauma his wrist has suffered from all of the high fives." Shortly before he won the presidency, an audio clip surfaced of Donald Trump responding to a similar case of a female teacher molesting a middle school boy with, "He might have put the moves on her. It might have given him confidence, actually."

One of the more heartrending conversations I had was with Alan, a senior at a Boston-area high school. While describing his most recent girlfriend, a classmate whom he said belittled and punched him, he mentioned, almost as an aside, that when he was thirteen, an eighth-grade teacher in her late thirties had coerced him into a seven-month-long sexual relationship. By the time I met Alan, she was serving prison time for child sexual abuse. He, meanwhile, had struggled with alcoholism, self-harm, anxiety, bouts of rage, and other abusive relationships. Would he have had those issues regardless? I can't say, but they are common enough in the wake of molestation or assault. Alan credits four years of intensive therapy with giving him some hope that he can move forward; he was lucky, he added, to have had access to such care.

Our culturally dictated ideas about gender, sex, and desire shape our vision of what assault looks like and who experiences it, sometimes dangerously so: as many as one in six boys will be sexually abused or assaulted before turning eighteen, yet parental concern focuses largely on girls. At its most extreme, sexual abuse of boys by adult men—the scandals in the Catholic Church, at Penn State, at Ohio State, at elite private schools, by celebrities such as Michael Jackson and Kevin Spacey—can remain hidden in plain sight for years, its victims having no language or feeling too ashamed to describe their experience. (While large-scale violation happens to girls as well—USA Gymnastics national team doctor Larry Nassar and University of Southern California gynecologist George Tyndall each abused hundreds—there is, for better or worse, some collective understanding of the potential risk.)

Acknowledging assault by another boy is equally taboo. A freshman at Brown who went public in an interview with the *Huffington Post* after being assaulted by another student in a dorm bathroom initially joked to friends he'd had a "five a.m. hookup": that felt easier than reckoning with his feelings of humiliation and loss of control. One young man I met, a sophomore at a Big Ten university, wanted to talk to me specifically because two other boys, brothers whom he thought of as friends, had molested him from age five through twelve. He hadn't wanted them to touch him and didn't enjoy it, but he also never told them to stop. Eight years later, those interactions haunted him. Although he'd gone through rehab in high school for excessive cannabis use, he was still dabbing heavily in an attempt to stop intrusive thoughts; his dorm room was littered with drug paraphernalia, liquor bottles, and boxes of half-eaten fast food.

He asked me whether I thought what had happened to him was "normal," whether it meant he was gay (he was not sexually attracted to men but hadn't been able to sustain a relationship with a woman), whether he was to blame. I was the first person in whom he had ever confided.

The SHIFT study found that students who experience assault—regardless of gender or sexual orientation—will, in conversations with peers, downgrade the incidents, labeling them "weird" or "awkward" or "regrettable" in order to keep the peace among friend groups or student organizations; for the sake of their own well-being; or because it doesn't fit their self-image. That may be especially true for boys, who in both Jessie Ford's interviews and the SHIFT research would make a joke out of unwanted sex, calling it "funny" or—here it is again—"hilarious," especially if their friends had found out about it. One male SHIFT subject talked about a woman who encouraged him to keep drinking, footing his bill at a bar. Although he wasn't attracted to her, he went back to her room and the two had intercourse. The next day, a friend framed the encounter as humorous, saying, "Dude, she was trying so hard to get you drunk." Neither of them labeled the encounter as assault. "There is this notion of gender and sexual scripts in which men in heterosexual sex are understood to be responsible for moving the ball down the field," observed Jennifer Hirsh, codirector of the SHIFT study. "Men are supposed to be the aggressors and girls the blockers, so that makes it hard for men to understand and label their own experiences of unwanted sex. Agency and consent are assumed. It also makes it hard for women to understand the necessity of *getting* consent from men."

Certainly the label of "victim" conflicts with notions of conventional masculinity, including perpetual sexual readiness, but the inability to recognize or process negative experiences ultimately robs boys of choice and, potentially, of empathy. That may explain something that caught my attention: while girls I'd met often sought more intimacy, trust, and safety in their personal relationships after a nonconsensual encounter, boys did the opposite, sometimes to the point of hostility toward women. Dylan began identifying with the rapper Eminem, whom he described as "super, incredibly misogynistic and violent toward women. . . . I felt like I kind of understood it." He vowed never again to take an emotional risk, never to expose himself to pain or betrayal. "I told myself that I would never love a girl again," he said, "I'd just hook up with them from now on. That's it." He only felt safe if he was untouchable.

Liam, the hookup king from the previous chapter, also had an unwelcome experience the first time he had intercourse, again with an older girl with whom he had made out once or twice before. She texted him out of the blue one day, saying she was "horny" and asking if he was up for a "booty call." Liam knocked back a couple of shots to calm his nerves and went for it. The sex felt good, he said—but it also *didn't* feel good. "I felt like I was being used," he recalled. "I was just there for her to have sex with, not as a person, which maybe I deserved because of my reputation. And it was still status, right? It was something I could talk about with other guys—and I did, but I never told them I wasn't comfortable with it. You can never say that as a guy. It's got to be great. It's not even 'It's got to be great,' it *is* great. Period. End of story. But it's made me more cautious, more nervous about intimacy and relationships. I think that's why I've never officially dated someone. I don't trust people as much, and I don't ever want to be that vulnerable again."

"I Don't Want to Deny It. That's What Happened."

For months, Dylan avoided girls—even platonic friends. He eventually confided in a couple of guys, who were surprisingly sympathetic, though they couldn't exactly relate. One said he wanted "to kick Julia's ass"; that didn't seem quite appropriate to Dylan. "It wouldn't fix anything," he said. "Although having to see her in the hallway every day is kind of heartbreaking, honestly." When he did try dating again, Dylan's new girlfriend pushed him to have sex. He kept putting her off without saying why. One night, at another party, she got drunk and started yelling at him, "Why are you being such a prude?" So he told her the truth.

"Fuck you," she responded. "Guys can't get assaulted for real." She apologized the next day, but the damage was done. Dylan blocked her on his phone and fell into a funk.

He met his current girlfriend a few months later, at another party. They chatted on a couch for a while, then she suggested they go someplace private; they kissed a little bit, but mostly they just kept talking. Dylan asked if she would like to hang out again and she agreed. She also accepted without question his request to take their physical relationship slowly. When they did eventually have intercourse, it was exactly as he'd hoped his first time would be; that devastated him all over again. "I'd talked with a few friends at that point who were like, 'Dude, virginity's a social construct; you don't have to count that other time.' But I also don't want to tell myself that that wasn't the first time, because it was. That's what happened."

When he confided in his girlfriend, about six months into their relationship, he began to sob, something he hadn't done, publicly or privately, for at least five years. He totally lost control. It was embarrassing, weak—and a relief. "It was like I exploded," he said. "It was really ugly, violent crying." His girlfriend held him and wept with him; it was at that moment that he started to feel a little better.

Dylan still sees the girl who assaulted him every day at school, though they never speak. "She was my friend," he said. "I trusted her. I wish I could tell her that I hate her with all my heart. I wish that there was a word stronger than 'hate' that I could use. I hope to God I never see her again after high school is over.

"I do think I know how to deal with really hard stuff now, though," he added. "It was a shitty thing, but I can definitely feel love again, and I learned that life goes up and down. It oscillates. I'll never be *always* happy or *always* sad, which has been the greatest lesson. Things may not be okay right now or even for a couple months, but they will be okay."

I asked Dylan one last question: What would've happened if the genders were swapped, if she'd been drunk that night and he'd been the sober aggressor? He laughed, but without amusement. "Yeah," he said, "I'd be expelled. I'd be in jail right now. Because it was textbook, right? I was basically unconscious. I didn't want to do it. There was no consent."

CHAPTER 8

A Better Man

Throughout this book, I have told you about boys. I have written about their understanding of themselves as men and as sexual beings: about their insecurities, their hopes, their occasional breakthroughs, and their flaws. I have described the harm they have caused others, whether out of thoughtlessness, recklessness, indifference, ignorance, or malicious intent. Sometimes harm had been done to them as well. And through it all I have wondered: How can we raise boys to be better men? How can we ensure that they see women and girls as full human beings worthy of dignity, empathy, and value in intimate encounters? How can we reduce sexual violence, and, when it does happen, how can we create appropriate accountability? What do we do with boys like Liam in chapter six, who was afraid that admitting to possible sexual misconduct would make him a pariah? Or Jackson, a high school senior who, while drunk, sent pictures to his friends of a girl giving him a blow job? When the incident went public, his female classmates shunned him, folding their arms when he crossed the stage at graduation and banishing him from the year's final party. He ended up feeling aggrieved, like *he* was the victim. Or Trent, a junior at a West Coast college, who was banned from parties after being anonymously placed on a list of alleged campus assailants? He left school for a semester (without telling his parents why) to commit himself to reading books on feminism, sexuality, and sexual violence; an accomplished musician, he also decided to give up performing to "make more space" for women and other underrepresented groups. Despite all that, he faced hostility on returning to campus and more than once contemplated suicide. And what about Darren, a fraternity brother and a junior, also at a West Coast college, who "stealthed" a freshman girl, secretly removing his condom during intercourse and ejaculating inside of her? The lack of basic sexual ethics revealed by the cascade of #MeToo allegations is clearly well established at younger ages, and, just as in the adult world, colleges (and high schools) have struggled with how to address it: survivors of assault can feel invalidated by school bureaucracy; perpetrators might be punished yet remain resentful, never truly acknowledging the hurt they caused.

That's, of course, when formal complaints are filed at all. A national survey of students on twenty-seven campuses found that even in the most extreme cases of forced penetration, only 28 percent of incidents—and only 13 percent of those involving penetration while incapacitated—are ever reported. Meanwhile, mothers of boys sanctioned for sexual misconduct, each of whom fervently believes her son was wrongly accused and denied due process, have spearheaded a high-profile backlash against anti-assault efforts. Their agenda, however, seems less about equitable reform than rollback. One woman, whose son was expelled after having sex with an incapacitated girl, complained to the *New York Times*, "In my generation, what these girls are going through was never considered assault. It was considered, 'I was stupid and I got embarrassed.'" (To which one might answer: true, and fortunately standards change; as recently as the 1970s, marital rape was also not considered a thing.) Another, who was among a delegation that met with US Secretary of Education Betsy DeVos, apparently believed that "You need to keep it zippered" was sufficient sex education (then again, DeVos, an abstinence-only proponent, would likely agree). The prevailing system, it seems, often fails to result in understanding, healing, or justice. Where was the story, the idea, that might move the discussion forward?

Then, through Stephanie Lepp, a colleague who hosts the podcast *Reckonings*, I met Anwen and Sameer.

"Get Some!"

Sameer grew up in Los Angeles, the only boy among five children. He was raised mostly by his mom, who was from El Salvador, though he often volunteered at hospitals or food banks with his father, who was Pakistani and Muslim. Mostly, like a lot of guys, he spent his time playing video games or sports. No one ever talked to him about sex or relationships. "The majority of my sex education came from LA public schools," he said, "and whatever I could pick up on the internet and through media: so your standard porn, your teen-geared thrillers, comedies, action films, TV shows, whatever." On his eighteenth birthday, his maternal uncles and older male cousins initiated him into the rites—and rights—of manhood by taking him clubbing. They told him if he saw a "hot" girl he should "just go up and start dancing on her. She'll turn around and look at you, and if she likes you, she'll keep dancing. If she doesn't, she'll walk away." According to them, he should be trying to "bang each and every piece of ass I could."

Sameer attributes some of his attitude to being the "child of two macho cultures," but his perspective didn't sound much different from that of a lot of guys I talked to, regardless of ethnicity. Sexual conquest was the measure of his manhood. An evening out wasn't complete unless he got someone's number. His persistence made girls so uncomfortable that some female high school classmates avoided him. Once, junior year, he badgered his girlfriend—whom he cared about deeply—until she got out of his car, slammed the door, and walked to a friend's house. "I knew I'd done something wrong," he said, "but I didn't really understand *why* it was wrong and she didn't have the words for it. So I was like, 'Why isn't she into this? What's wrong with *her*?' That mangled that relationship really quick."

Even after the mandatory consent assembly during orientation at the midsize college he attended in the Pacific Northwest, Sameer thought of rapists as strangers who jumped out of dark alleys; that, or the scumbag who roofied one of his sisters at a holiday party. "So it was like, 'All right, cool,'" he said. "'I will be a bystander, and if anybody looks creepy or a girl is uncomfortable, I will be a hero, save the day, do the right thing!' That's about all I got from consent education."

Sameer was introduced to Anwen a couple of months later at an off-campus party. She was from a small northwest city, passionate about partner dancing—Lindy Hop, blues, tango. She even made her own skirts, designed to flare out as she spun. A short, slight girl, she was struck by how tall and broad-shouldered Sameer was, dapper in his gray shirt and black tie. He was taken by her eyes, her smile. "I was like, 'Wow! This person's gorgeous,'" he said. "'I'd like to get to know her better.'" She agreed to dance with him as long as they didn't grind. Sameer claimed to know some swing moves. "She then proceeded to humor me for the next five minutes while I tried," he recalled.

A few days later, he tracked her down on social media and asked her out. She hedged, not sure that she was interested, but not convinced she wasn't. "Well, we can just hang out," she finally said. They planned to go bowling, but all the lanes were full, so they bought ice cream at a local market—Anwen insisted on paying for hers, to show that it wasn't a real date—and sat outside, talking. Both remembered enjoying the conversation, chatting and laughing for several hours. Another time, they went for a walk together. Sameer dropped Anwen back at her dorm, hugged her goodbye, then leaned in for a kiss—nothing too heavy, but enough for Anwen, who had only ever kissed one other guy, to realize the chemistry wasn't there. Sameer continued to text, showed up at a couple of partner dance events, insisted on bringing Anwen soup when she was sick. She responded coolly. Eventually, she stopped texting back, and, figuring she wasn't interested, he let it go.

That spring, they both went Greek. Anwen hadn't been interested in sororities, but all her friends were rushing, so she did, too. Sameer, who'd been a fan of the National Lampoon *Van Wilder* movies in high school, was eager to pledge. "I thought college was a place to, yeah, learn a lot, but also to party hard," he said. All fall, he'd walked past the fraternity he hoped to join, seen the girls heading inside in their skimpy outfits, heard the music blasting. "It seemed like the coolest thing ever," he said. What clinched it, though, was a conversation he had with a senior brother about their service work, the chance to give back to the community. That appealed to Sameer, who missed the volunteering he'd done with his dad. "Later, I found out that was very much a sham," he said. "Mostly, frat life was about binge-drinking and hooking up." Like a lot of guys I met, Sameer was quick to tell me that his fraternity was "not the worst," not the "bro-iest" or the "frattiest," that other houses had the overt reputation for assault. "But then," he added, "that's kind of like saying we were the least deadly of all the sharks, isn't it? Like, 'Yeah, we only kill *one* person a week!' But you're still killing people. So what the fuck?"

Anwen knew Sameer would attend the big party welcoming Greek recruits. She even joked to her friends as they were getting ready not to let her go home with him. The two quickly bumped into each other heading opposite directions on a staircase, just after Anwen had dumped her jacket in a brother's room that was being used as a coat closet. "Hey," she said. "I just want to apologize for completely stopping texting you."

"Don't worry about it," he said. "I understand. You're not looking for a relationship."

No, she thought, *it's more like I'm not looking for a relationship with* you. But of course she didn't say that out loud. Instead, she smiled and headed for the basement to dance with her friends. Sameer eventually joined them, and they began dancing together. Face-to-face swing turned to blues, but as that turned to grinding and then kissing up against a wall, Anwen grew uneasy. Sameer grabbed her hand and led her to an adjacent room, where he pulled her onto his lap. Being held in someone's arms felt good, but Anwen didn't want that someone to be Sameer, and she wasn't sure how to gracefully disengage. She caught the eye of a guy she knew, hoping he'd sense her distress and intervene. He didn't. Eventually, she told Sameer she needed to get her things and go home. But it turned out that the room where she'd left her jacket, along with her phone, her ID, and the key card for her dorm—basically everything except a tube of ChapStick—was locked, its occupant nowhere to be found. Sameer got the guy's number from someone and tried to call him, but the phone went straight to voice mail. They searched the house to see if he'd stashed Anwen's things someplace else: nothing. What's more, her friends, the ones she'd asked to keep her from leaving with Sameer, had vanished.

Sameer thought this was his chance: since she was stranded, she could stay in his room, he suggested, but Anwen declined. He, characteristically, pushed. "I remember pulling a classic 'soft boy' move," he told me. "Like, 'Fine. Leave me alone. Everyone always does.' Basically, 'Oh, woe is me, my life is hard, you should feel bad for me, don't leave me.' Very emotionally manipulative shit. And it worked." It was late. It was cold. Anwen didn't know what else to do—so she eventually said okay. As they left the frat, she spotted a couple of guys from her dorm down the street and told Sameer thanks, but never mind: she would run and follow them home. Again he wasn't giving up so easily. "You can't just leave after kissing me like that," he said. She gazed after the retreating boys, calculating whether she could actually catch them. If she didn't, then what? So, believing she was taking charge of her situation, she turned to face Sameer. "Okay," she said. "Let's go talk about this." They went inside an academic building that was open all night; Anwen could hear people whistling at them, calling out, "Get some!" They sat on a couch in a lounge and she tried to explain to Sameer that she wasn't interested in dating him, she wasn't interested in *anything*; at the same time, she was trying not to hurt his feelings, to let him down easy, to tell him he was a nice guy and she'd had fun dancing with him. It all came out kind of muddled. One point she was very clear on, though: "I don't want to have sex with you."

"That's okay," he replied. "I don't have any condoms." Which was not, of course, an actual response to what she'd said.

They started to kiss again, and Sameer, still very drunk, pushed Anwen down on the couch. "I can make you feel good," he said, covering his body with hers.

"He was so much larger than I am," Anwen recalled. "I was totally trapped. That was the moment I started to panic a little bit. So I said, 'No, Not here.'" He led her into the men's room, pinned her to a wall Hollywood-style, and began kissing her again. Again she said, "Not here."

Sameer phoned his roommate, who begrudgingly cleared out—they had a rule against bringing girls home after two a.m., but Sameer pleaded, "Come on, just this once, man!"—taking his bedding with him. When they got to his dorm, Anwen removed the shorts she'd worn to the party, leaving on her leggings and tank top. She popped her contact lenses into a shot glass filled with water (never a good idea). Then she went to the bathroom. Sitting in a stall, she racked her brain for any friends she had in common with Sameer, anyone whose number might be in his contacts. *What are you doing?* she thought. *What are you doing?* Maybe she could sleep in the dorm's common room. "But I knew he'd say that was silly: there was a bed in his room, blankets." Her mind reeled. It was so late. She was so tired. She just wanted to sleep.

Looking back now, Anwen wishes she had thought to go to campus security. She wishes she had followed the two boys back to her dorm. She wishes her friends hadn't abandoned her. She wishes the fraternity brother who locked her stuff in his room had told her that he was leaving. She wishes she lived in an era where people still memorized their friends' phone numbers. "*So many things*," she told me. "*So many things.*" But none of that is what happened. Chance and choice colluded: people don't always make the decisions that seem patently obvious in hindsight.

When she returned to Sameer's room, he had queued up Bon Iver's "Skinny Love," which Anwen had said was one of her favorite songs. He began kissing her against the closed door, scooped her onto his bed, then climbed up after her. He knew her sexual experience was limited; his was, too, but the girls he'd been with before "liked it rough," or at least appeared to, so he began rubbing Anwen between the legs, hard. It hurt.

"Oh, fuck," he said, "it would feel so good to fuck you."

"Remember, I don't want to have sex," she replied.

"It's okay, I don't have a condom," Sameer said again.

A few minutes later he told her, "Take off your shirt"—more an order than a request. She had pulled her top halfway off when he grabbed a breast and squeezed. A wave of revulsion hit her, and she jerked away, blurting, "No!"

"It's okay," he said. "That can come later." Sameer figured he'd try again when Anwen was more comfortable: he attributed her hesitation to nerves and inexperience. He imagined he would be "the nice guy," a "teacher" who would help her along without judgment. He took her hand, guided it to the crotch of his sweatpants, and began to rub, then he pulled out his penis. "You should play with it," he said. "It doesn't bite."

He continued holding his hand over hers. "Wow, you really haven't done this before, have you?" he commented. Then added, "Your mouth would feel even better."

He put a hand on top of her head and, Anwen recalled, "pushed down, pushed down, pushed down. This was not something I'd ever done before. And it didn't feel good. I didn't want to be doing it. It felt like I was going to choke. I was gagging."

Thinking that, as a newbie, she was growing frustrated with the "tips and tricks" he was offering, Sameer eventually let her up. "I know I'm hard to please," he said, believing he was being magnanimous. He got up to go to the bathroom. "Let me just go finish up."

Anwen remembered thinking: *I didn't want to* please *you. I didn't fucking want to* please *you.*

He came back and kissed her good night. For several hours, she lay in bed, scooched as close to the edge as possible, Sameer's arm flung across her stomach, her eyes fixed longingly on the roommate's stripped mattress. She cried quietly. And then she slept.

As far as Sameer was concerned, the night had been a success. He lent Anwen a sweatshirt the next morning, secretly thinking that if she kept it, he'd have an excuse to see her again.

"It's so big on me!" she said.

He laughed. "*I'm* so big on you!"

He walked her to her dorm, kissed her once more. When someone opened the locked door to leave the building, Anwen grabbed it and dashed inside. "So, how was *your* night?" her roommate teased. Anwen was vague. A couple of days later, she told friends that she'd spent the night with Sameer, that the two of them had made out. "And," she remembered, "I really didn't know, so I was like, 'Do I have to start dating him now?'"

They got together once more, ostensibly to study. He leaned her back against a table and kissed her, pressing a knee between her legs. "You like that, don't you?" he said. She did not, and wriggled away. After that, she went back to ignoring his texts. Sameer was disappointed but philosophical: "Fairly often at our school, people stopped talking after a hookup because they felt awkward and didn't know how to communicate." He didn't grasp that he'd done anything amiss, and Anwen wasn't about to tell him. In fact, she avoided him completely, leaving a room if she happened to spot him.

Still, she couldn't get that night out of her head: the images popped up unprompted, causing her to panic and gag, especially if she read or saw anything involving sex or assault. Always a straight-A student, she couldn't focus on her studies and had to drop a class she was in danger of failing. When, sometime later, she began dating someone, physical intimacy, particularly oral sex, felt fraught.

Up until this point, Anwen and Sameer's story was relentlessly, depressingly ordinary, the kind of episode that happens at high school parties and college campuses every Saturday night. Then, at the beginning of his sophomore year, as part of training to become a student orientation leader, Sameer attended a presentation by a representative of Green Dot, a program that instructs students on how to intervene and defuse situations that could potentially lead to "power-based violence." "The person was talking about how assault wasn't only physical force," Sameer recalled. "It could be emotional manipulation and coercing someone into a sexual act. It could be putting someone in a situation where they didn't feel like they could say no. And right away, I remembered that night. I thought, *Did I do that? Does Anwen see it that way? If she does, why hasn't she reported me?* I was terrified. I was terrified that I'd assaulted her. I was terrified that I'd hurt her. I was terrified of what would happen if she reported it. And I was terrified because, if this was true, then who was I?"

Much of his immediate reaction, Sameer said, centered on what could happen to *him* if Anwen filed a complaint—expulsion, maybe jail—but over the next year, he began to engage in some serious soul-searching. He attended another, longer Green Dot training and successfully lobbied to make the program mandatory for new recruits to his frat. He read whatever he could find about consent and healthy sexual interactions. Eventually, he took courses on human sexuality and romantic relationships. He talked to his female friends about their experiences, too. He began to recognize that most assault didn't happen between strangers. In retrospect, he was also pretty sure Anwen was not the only girl he'd harmed.

The following fall, Anwen, too, became an orientation leader. One night, as Sameer was walking back from an event he had staffed, he heard her call his name from the shadows. It was the first time they'd spoken in over a year. He turned toward her, his heart pounding.

"Can we talk?" she said.

"Sure." They went to a quiet spot and sat on the ground, chatted uncomfortably for a few minutes.

Then, "I want to talk about that night," she said.

"Just to clarify, you mean the night you came home with me?"

She nodded. "Yes. I want you to name it."

Sameer hugged his knees to his chest. He rocked back and forth. He tried to speak, but nothing came out. He tried again and croaked. "Rape."

Relief flooded Anwen. She wasn't crazy. This had really happened. But she said, "I wouldn't call it rape"—although she'd later learn that, according to their school's conduct code, it was—"I'd call it sexual assault."

A moment later, she added, "I forgive you."

"If it's okay with you," he responded, "I don't think I'll forgive myself just yet." Sameer told Anwen all he'd been doing to educate himself. He offered to turn himself in to the police, but she said she didn't want that. She just wanted them to keep talking, for him to understand how the night had affected her.

"Of course," he agreed, "whatever you need."

So periodically, Anwen would text Sameer and they'd meet. "I learned that the reason I hadn't seen her for so long was because she had systematically memorized my schedule so she could take routes to avoid me. Because the mere sight of me could ruin her emotionally for the rest of the day." They met a handful of times, but talking to Sameer face-to-face was too much for Anwen; it overwhelmed her. When she left campus for a semester abroad, yet again, she cut off contact.

The next time they saw one another, the fall of their senior year, was at a campus Take Back the Night march. Anwen went as a survivor of assault; Sameer was supporting his new girlfriend, who had her own history of sexual victimization. "It was strange," he said, "because I was very angry at the person who had done this to her, to this woman I really loved, and yet I was responsible for similar acts." Sexual activity could send his girlfriend spinning into PTSD. "Something would happen and her eyes would glaze over and she would be right back with her rapist. I had to get really good at noticing and zeroing in: stopping everything and trying my best to use my voice and physically be there with her to bring her out of that nightmare and back into reality." Sameer learned to be meticulous about consent: move slowly, carefully, and check in frequently to be sure she was not only okay but enthusiastic.

Sameer worried that he didn't deserve to be at the march, that his mere presence was a form of violation, but his girlfriend, who was aware of his history, said she needed him there, so he went. The evening culminated with an open mic, and Anwen spontaneously rose to tell her story. She locked eyes briefly with Sameer, who was sitting just a few feet away. "I was absolutely panicked," Sameer recalled. "All I could think was, *Shit! Oh my God! How the fuck?* I had no idea what she was going to say. She talked—not in detail—about what had happened to her, and one of the last things she said was, 'If the person who did this to me comes out and tells his story, I hope you'll listen.'"

What she meant by that was, she hoped if Sameer could muster the mettle to take public responsibility for his actions that the group could hear him, could acknowledge his humanity even as they condemned his crime. Sameer was floored by her generosity. Even so, that night he remained silent.

Around the same time, one of Anwen's closest friends filed a formal complaint against a guy who had, essentially, done the same thing as Sameer: the only difference was the girl had been incapacitated at the time, while Anwen had not. She asked Anwen to accompany her to a hearing where both students, as well as witnesses, would give statements and be questioned by a three-person panel of specially trained faculty and staff. The committee found the boy responsible for misconduct, but not much came of it. He was briefly suspended, possibly faced some other minor sanctions. The girl was satisfied—she had just wanted *something* to happen—but Anwen suspected the guy, who until then had been a friend of hers as well, never really believed that he'd done anything wrong.

Going through the process, meeting the staff at the office of student conduct, and booking a few sessions with a therapist friend of her mom's made Anwen realize that she, too, wanted to file a formal complaint, but she was put off by the standard disciplinary proceeding in which neither the accuser nor the accused had a voice in the outcome. The assault had made her feel powerless; she didn't want the resolution to do the same. "My friend wanted a higher authority to tell the guy he'd done something wrong and what he had to do because of that," she said. "That's totally valid. But I needed to be integrally involved in creating something that wouldn't just fix our issue but could also maybe open a couple more people's minds." She wasn't looking to have Sameer suspended or expelled. She wanted him to be actively involved in deciding how he could make amends. She wanted to be an agent of change, not punishment. "I didn't want his power taken away, either," she continued. "The current system creates resentment because the verdict is just handed down by someone. There's never a point where you get to understand the other person as a human being. I believe we have such potential for compassion and understanding, but you have to *talk* to each other. You have to hear that firsthand experience."

From "Monster" to Man

The school's director of student conduct, Frank Cirioni, had long been interested in what are called "restorative justice (RJ) practices," though he'd never personally used them for a sexual misconduct case. Unlike conventional disciplinary proceedings, in which a disinterested panel determines whether the accused has broken a rule then metes out punishment accordingly, restorative justice (a term encompassing a range of interventions) seeks to "repair harm," to the extent that is possible. Trained facilitators guide everyone involved—those who have been hurt, whether an individual or a community, as well as those responsible for damage—through a multitiered, collaborative process in which they describe the incident, its impact, what needs have been created, and what obligations and engagement should result. Although its primary emphasis is on healing those wronged, RJ aims to be transformational for everyone involved by creating true accountability and reducing the risk of recidivism. Restorative practices have been used around the world to address issues as diverse as juvenile crime, murder, genocide, and the aftermath of civil war; they are already used on campuses for other violations such as cheating and underage drinking. In his study of 659 conduct cases at eighteen colleges and universities, David Karp, a sociologist and director of the University of San Diego Center for Restorative Justice, found that students engaging in restorative versus retributive procedures were more likely to say they had a voice in the proceedings, to feel that they'd taken responsibility for their actions, that the process was fair, and that they were ready to move on. They also felt stronger ties to their school communities.

RJ is not a magic bullet. "Respondents," as they are usually called, must, from the beginning, be willing to admit fault, and there will always be those who refuse, or who are indifferent to the impact of their actions, or who aren't capable of moral engagement. Nor can RJ replace all other forms of justice, though it may reduce their application. As with conventional hearings, it may fail in its purpose, requiring additional measures as a backup. There's a risk that victims who might, in fact, want a process with a more potentially punitive outcome (such as expulsion) could feel inappropriately pressured to forgo that route by administrators or peers. What's more, as Judith Lewis Herman, a professor of psychiatry at Harvard University Medical School, has pointed out, without feminist-forward leadership, the "community standards" guiding RJ could reproduce traditional biases against victims of sexual violence. There is no perfect system, but

RJ does offer an important alternative to current campus adjudications that so often leave survivors retraumatized and offenders hostile. Ideally, its attention to victims' needs, to education, and to strengthening communities would increase willingness to report assault, expand the proportion of offenders who are held accountable for their actions, and inspire authentic cultural change. "My mantra," Karp said, "is 'What are the conditions in which it's possible for students to admit responsibility for the harm they've caused?' The systems we've put in place do the opposite. They put these guys in a position where it's only rational for them to deny responsibility, or to minimize or displace it. They hear from their parents, from their lawyers, from Brett Kavanaugh, from everywhere that that's the only thing they can do to protect themselves from this terrible accusation of being a sex offender. We're creating a pathway to acknowledging and hearing the harm they've caused. That is absolutely the goal of the process."

Cirioni had a personal stake in restorative practices. As an undergraduate RA in the early 2000s, he had sexually harassed a female colleague for over six months, pestering her to go out with him, even more so after she'd dated one of his fraternity brothers (*Why him and not me?*). She eventually reported him, but, while he was found responsible for misconduct, there was neither reprimand nor consequence; he didn't even lose his RA job. More important, he learned nothing from the experience. A few years later, as a graduate assistant, he harassed another, younger student, expressing sexual interest, touching her inappropriately, trying to convince her to meet him outside of class. She, too, reported him, and while again, he faced no meaningful penalty, he was required to read a formal statement she'd written describing the impact of his actions on her. That hit him in the gut. "It was the first time I could truly hear how I had made someone else feel," Cirioni recalled. "It's when my life really changed, and may be why I found myself learning about restorative justice a few years later and thinking, *This feels like a way to connect.* You have a better chance of making change when you say to someone, 'We're not going to exclude you or remove you; we're going to help you. You are not a bad person, but your behaviors are concerning, and we need to examine those and where they come from and why you engage in them.'

"We're all flawed humans," he continued. "We've all caused harm. Some wells are definitely deeper than others. The real question is, how do we have a chance to take responsibility and make things right?"

If Sameer's initial behavior toward Anwen was a perfect storm of gender socialization and ignorance, his actions in the aftermath—along with Anwen's self-awareness and desire for an alternative process—made Cirioni believe they would be an ideal test case for applying restorative practices to sexual misconduct. "When I told Sameer that a report had been filed against him," Cirioni told me, "he immediately said, 'I know what you're talking about, and I take full responsibility.'"

Over the course of a semester, Cirioni met separately with Anwen and Sameer on multiple occasions in what is called a "preconference": educating them about restorative practices, listening to their narratives of the incident, preparing them to meet. Anwen had previously told Sameer how profoundly that night had affected her, but now she needed him to know specific details. She wanted him to read everything she'd written about it: poems, essays, journal entries, her formal "impact statement." It was the hardest thing Sameer has ever done. "I saw that what I had thought was an 'awkward hookup' where I was trying to be kind of a teacher was, for her, an account of fear and discomfort and violation and—just *pain.* In my mind, I had always tried to be someone who did right by people and treated them with love and respect. But this guy who forced himself on this girl? That was me.

"It was hard not to view myself as a monster," he continued. "That's the biggest word I can use. How could I have made a person like Anwen, who is so nice and so kind, feel this way? I genuinely hated myself. And beyond that—I grieved. I felt fucking awful. And anything I could do to make her life better and easier—*done.* Absolutely."

Anwen wanted a letter of apology, but she didn't need Sameer's guilt. She didn't need his shame. She didn't need his acquiescence to her every request. She needed him to act, to think, to come up with his own ideas for making things right, or as right as they could be. Eventually, they developed a plan. Sameer would tell his story publicly: he wrote an article for a campus magazine (signing his real name); he cowrote a spoken-word piece with Anwen that they performed together at a Green Dot training, which, in part because of Sameer's efforts, was now mandatory for every recruit to Greek life; he talked to me for this book. Sameer would also strive to educate other men about consent and assault. He met with officials from local high schools, hoping he could talk to boys to show them that someone who perpetrates assault could be an ordinary guy—to encourage them to make better choices; to keep them from having to learn, as he did, at the expense of someone else's suffering. No one, though, was eager to have an admitted assailant address their students. I think that's too bad, as Sameer is exactly the kind of guy that young men need to hear from—someone just like them, someone just like they *could be,* for better or worse. "Nights like the one with Anwen are so common," Sameer said. "That's how guys learn to operate in a lot of ways; our level of understanding of how to communicate and navigate sexual relationships is so infinitesimally small. They don't have the frame of reference to understand what it means to be a good partner, a good lover. So a lot of us are guilty of doing things like this, and we need to start talking about it and owning up to it."

Sameer also started talking more directly to his male friends, challenging their hookup narratives. "They'd be like, 'I hooked up with this girl! It was great!' And I'd ask whether *she* enjoyed herself. Guys are taken aback by that response. I'd be like, 'Did you ask her?' And they'd either be silent or say that it would be too weird. But why is it weird?

"I got into the habit, and I'll say this to my guy friends, of doing a kind of—debrief, I guess, with my partner. Like, 'Hey, what did you like? What didn't you like? What might you like to try?' Just the standard conversation that needs to happen or else people will just keep having bad sex and faking orgasms and lying to each other about what makes them happy sexually."

As part of their process, Anwen and Sameer also met once together, with Cirioni carefully facilitating. Even though they had spoken before she filed her complaint, seeing Sameer in an official setting made Anwen anxious. Sameer was nervous, too, unsure of how to behave. Anwen wanted answers: *Why did you do it? Didn't you see I was panicked? How do I know you'll never do this again?* Sameer talked about the reading and thinking he had done, the conversations he'd had, the classes he'd taken: "If all this work and awareness doesn't stick," he told her, "then there's no help for me."

Sometimes in our conversations, Sameer would refer to his past self as "younger Sameer" or "freshman Sameer." I understood that. He was such a different guy now—so reflective, so empathic, so conscious of others' feelings and comfort. His arc seemed almost too good to be true; I was tempted to write him off as a unicorn. On the other hand, there was nothing about "young Sameer" that hinted such transformation was possible. He had been a regular guy, someone who'd absorbed regular guy ideas, who'd behaved like so many regular guys do. He was not exceptional; if he could change so profoundly, maybe others could as well. "When you realize that you've done something terrible, you're terrified of being judged and ostracized by your friends—to be fair, rightly so. That's why I'm such a big proponent of restorative justice: I want to believe that people, men especially, have the capability of being kind, empathetic, overall good humans who, if we're told that we're doing something wrong, have the ability to step up."

Anwen and Sameer had graduated college by the time we spoke. He was back in California, working as a bartender; she was living in a woodstove-heated cottage in the Pacific Northwest. They still checked in with each other every month or so and expect they'll always be in touch; although it isn't the goal or expectation of restorative justice, they've come to share an unanticipated intimacy. "It's pretty cool," Anwen told me, "and it's taken years to get there. But I know the worst thing he's done, and he knows the thing that's hurt me basically the most in my life. So I'm pretty comfortable talking with him about other things that have been hard. Because I know that he's taken these steps to become a really understanding, caring, growing individual."

These days, when Sameer tells people his story, they often try to let him off the hook, saying what he did wasn't really "that bad." That rankles him. "This is not a competition," he said. "And also, what do you mean by 'wasn't that bad'? Forcing someone to do sexual things against her will, emotionally manipulating her, *wasn't that bad*? Perpetuating a culture that makes people feel like they can't say no or can't be themselves or makes them feel scared *wasn't that bad*? You're affecting people's lives. You don't think it's *that bad*? That's simply not a qualifier I'm willing to accept.

"It's weird, though," he continued. "Some people are so quick to come to a perpetrator's aid, to justify his actions and try to make him seem like he's not a bad guy. Then there's other folks who are like, 'Are you fucking kidding me? This person is actively garbage, a piece-of-shit monster.' But there's not really that middle ground. Going through the restorative justice process and talking with Anwen gave me an opportunity to view it in a different way. The end goal is to view myself as a person who has done bad things, not just as a bad person. That's a really hard thing to do. I don't know if I'll ever quite get there. At the very minimum, the silver lining becomes that you realize what you did was wrong and won't ever do it again, and maybe you become someone who actively works every day to be a better person. To be a better man."

CHAPTER 9

Deep Breath: Talking to Boys

Restorative justice may be promising, but it's still an after-the-fact solution. Until young people—girls as well as boys—are better educated about gender socialization, sexual consent, ethical engagement, mature relationships, and diverse orientations, we will be stuck in damage control mode. Such lessons ought to be actively integrated into school curricula, taught in, though also beyond, human development courses (which we colloquially, if too narrowly, call "sex education"); instilling basic values of citizenship is, after all, part of the job of educational institutions. Yet, realistically, only twenty-four states and the District of Columbia currently mandate sex education, and only ten require that it be medically accurate. You remember that health teacher from the movie *Mean Girls*, the one who warns, "You're going to want to take off your clothes and touch each other. But if you do touch each other, you *will* get chlamydia—and die"? Well, right this minute, American students are learning that HIV can be spread through sweat or tears, that women who have sex before marriage are like chewed pieces of gum or used pieces of tape or cups full of spit (i.e., no one will want them), that boys should seek out "good girls" who say no to sex, that condoms are useless against HIV a third of the time, that the pill has an 80 percent failure rate, that abortion causes cancer. I fear for those young people, I do, and for good reason: despite a federal investment in abstinence-only education to date of over two *billion* dollars (and counting), teenagers exposed to its lessons have been found to neither significantly delay intercourse compared to others nor have fewer partners; they do, however, have higher rates of pregnancy and STDs. Equally concerning, while pleasure-based sexuality education that includes practicing refusal skills has been found to reduce rates of assault, abstinence-only programs have not.

Unfortunately, so-called comprehensive sex ed is not necessarily much better, focused solely on averting disaster: avoiding pregnancy and preventing disease. Fewer than half of high schools and only a fifth of middle schools teach all of the sixteen topics the Centers for Disease Control considers to be essential, such as creating and sustaining respectful relationships; understanding the influence of media, peers, and family; and developing communication and sexual decision-making skills. Note that list does not include understanding consent; only ten states require that, if taught at all, sex education include *any* discussion of the issue. One of them is California, but in my local high school that has meant a single class period conducted in ninth grade by an outside educator; when I asked a group of my daughter's friends about its content a year later, none could remember any specifics. We have arrived at this precarious state because the forces that want to ban positive, thorough sexuality education—among them parents who place ideology over their children's health—are typically more vocal than its supporters.

That may reflect, in part, an ambivalence even among progressives, a lingering belief that talking to teenagers about sex gives them license to engage (and, conversely, that if we avoid the topic, they will somehow *not find out about it*). That, it should be obvious by now, is a myth: decades of research have made clear that talking to children about sex does not reduce the age at which they start. Our teens are in urgent need of high-quality human development courses. Until those exist, relying on school sex education is a risky bet. And that means unless caring adults step up—parents, physicians, youth advocates, faith leaders, coaches—the default educator *will* be the media; it is impossible for me to believe that we would be so cavalier, so indifferent to any other aspect of children's development that was so integral to their safety, futures, and well-being.

After nearly a decade of reporting on teenagers and sex, if I know anything for sure, it's that parents just have to get over it. I know it's awkward. I know it's excruciating. I know it's unclear how to begin. You may have never even been able to have such conversations with your own spouse or partner. I get that. But this is your chance to do better. Discomfort and embarrassment are not excuses to opt out of parenting (quick tip: talk during physical activity. Or, even better, in the car: you don't need to look at each other, plus they can't escape). Despite their eye-rolling, earplugging, and other superficial resistance, teenagers consistently say that they do want such information from parents, and that they benefit from it. I know from experience that's true: boys often told me that our conversations had dramatic, ongoing, sometimes therapeutic impact—and I was a total stranger. So, rather than fixating on how discussing physical and emotional intimacy makes you—and your son—want to sink into the earth, consider the opportunity it creates for a closer relationship, to show him that you are genuinely there for him, to display openness, strength, and perseverance in the face of messy realities. How, after all, will he be able to have those challenging conversations as an adult if you don't pave the way now?

In that spirit, here are a few thoughts to get you started.

It's Not "The Talk"

Just as a single "talk" about table manners wouldn't make your son polite, a single discussion about intimacy won't ensure good sexual etiquette—particularly since, for parents of sons, the average length of such talks is ten minutes. Parents need to have *habitual*, brief, often casual conversations that increase in complexity as children grow older. By now it should be abundantly clear that the content of those chats has to range beyond anatomy, reproduction, contraception, and disease protection to encompass what it means to be (and to have) a caring, respectful sexual and romantic partner. An article you happen to read on hookup culture, or the latest swipe apps, or teens sharing nudes can be great discussion fodder. Ask him how it relates to his friend group, what pressures he thinks guys and girls feel in various situations. What are the roles of privacy, vulnerability, intimacy? Whose needs are prioritized? What does coercion or pressure look like?

Consent Is Crucial

Most of us adults did not grow up with the current "yes means yes" standard; some of us even predate "no means no." If you find the rules confusing, educate yourself first, so *you* understand that consent should be affirmative (silence is not consent); knowing (a person cannot consent while asleep, involuntarily restrained, or incapacitated by drugs or alcohol); ongoing (saying yes to one activity does not imply consent to another, nor does having done something in the past grant permission in the present); revocable (you must stop *instantly* when it is withdrawn); freely given (not coerced, won over, or manipulated—pushing someone's head down is not consent). If you aren't crystal clear with your son about all of this, you are putting both him and his future partners at risk: at risk of being assaulted, at risk of assaulting, at risk of not understanding what he has done until it is too late (and maybe not even then).

Boys, specifically, should be warned that they are prone to overestimating female interest and enthusiasm for sexual activity, all the more so if they have been drinking. They should know that consent applies to digital images, too: if you receive a nude photo, you may not show it around, forward it, or post it without the permission of the person depicted (and if that person happens to be under eighteen, possession or distribution of the photo may, even if consensual, be in violation of child pornography laws). If your son does commit sexual misconduct, he should know that *because* you love him, you will hold him accountable. While you can support your boy's learning and growth, you cannot enable his denial, deflection, or avoidance of taking full responsibility for his behavior.

Today's generation should also be reminded that consent has to be obtained in person (not via text or social media). What's more, it's a two-way street: boys, too, must voluntarily agree to physical intimacy, either verbally or through body language (though recall that erections do not equal consent), and should be subject to neither manipulation nor pressure, whether from girls or other guys. They, too, have the right to say no to unwanted advances; parents can help their sons strategize on how to do so if they fear social backlash. And you can never say too many times that assault or molestation—whether by a male or female, a peer or an adult—is never the victim's fault.

But Sex Is Not *Just* About Consent . . .

Consent is imperative, but it is a baseline. As health educator Shafia Zaloom says, consent is what makes sex *legal*, but it doesn't make sex *ethical*, and it doesn't make sex *good*. Let's say a teenager consensually hooks up with his girlfriend's best friend. Criminal? No. Principled? Likely not. Zaloom, a high school teacher whose own book contains multiple real-life scenarios that adults can work through with teens, tells her students that *ethical* sex means taking into account the well-being of not only the participants, but others who might be affected by their actions. "Good" sex is not only legal and ethical, but pleasurable and mutually satisfying. For that to be the case, boys must have an accurate conception of female bodies and sexual response. If you can say the word "clitoris" out loud to your son, by all means go for it. If you'd rather poke yourself in the eye with a fork—and many of us would—make sure to provide ample, accurate educational resources (books, websites) on informed, progressive sexuality (for ideas, check out my website: https://www.peggyorenstein.com/positive-sexuality).

Along with sexual information, young people (of all genders) say they want more insight from their parents about emotional intimacy: how to begin a relationship, establish mature love, how to avoid being hurt (or perhaps how to accept the potential for growth in pain), how to manage conflict, and how to deal with breakups. You needn't have had unmitigated success in your own romantic life to drop some wisdom on all of that. "We fuss a lot about 'the sex talk,'" says Richard Weissbourd, the director of Making Caring Common. "We do not fuss about a far more important talk, which is: How do we talk to our children about the courage and subtlety and discipline and tenderness and tough-mindedness that it really takes to love someone else?"

In reality, according to Andrew Smiler, a psychologist specializing in adolescent male behavior, most guys prefer physical intimacy with someone they know, trust, and with whom they feel comfortable. In contrast to the "always down for it" image, Smiler found in his surveys of heterosexual high school students that boys' greatest motivator for pursuing sex was not physical, but emotional: expressing love or a desire to form a closer relationship with a girl. "Guys say they appreciate having a dating partner," he told me. "That special person who they know will have their backs no matter what, someone they can talk to about anything, especially the things they feel are off-limits with their male friends. They don't get that with a one-night stand or random hookups.

"As adults—whether parents or teachers or professionals—we might ask boys what *kind* of sexual experience they want," he continued. "Not just whether they are looking to have an orgasm, but about the context and quality of that orgasm. If we're willing to be more vulgar and pointed, we might even ask, Do you want a partner who's more than just someone to masturbate into?"

And It Isn't Only About Intercourse

By continually equating "sexually active" to penis-vagina intercourse, we allow kids to label other acts—including manual, oral, and anal sex—as "not sex" and so potentially not subject to the same rules; that opens the door to risky behavior and disrespect. It also turns sex into a race to a goal rather than a pool of experiences involving affection, warmth, closeness, desire, touch, arousal, intimacy. You might ask a teen to consider who is truly the more sexually "experienced" person: someone who has kissed a partner for three hours, experimenting with sensuality and communication, or a guy who gets wasted at a party and hooks up with a random girl in order to punch his V-Card before college? (Hint: it's not the latter.) Amy Schalet, a sociologist who studies adolescent sexuality, defines true sexual autonomy as learning to assert wants, set limits, and behave responsibly in sexual encounters. The best way to do that is to move slowly, with awareness and trust in a partner.

Broadening the view of sex also means being LGBTQ+ inclusive. Straight kids need to understand other sexualities to avoid demonizing or marginalizing them. If your son does identify as gay or bisexual, your family's acceptance, obviously, is vital, but not enough. As with any child, you still have to express your values around sexuality, intimacy, and relationships. You need to provide accurate information and resources on engaging in a variety of partnered practices, including anal sex, safely and enjoyably—if you don't, you are putting him in harm's way. It's critical to talk about PrEP (a daily pill that can prevent HIV in high-risk individuals) as well as other disease protection. And if you'd rather your sixteen-year-old didn't claim to be eighteen and go on Grindr behind your back, you may need to brainstorm about ways he can meet boys closer to his age (especially in high school or middle school) so that he has the same opportunities for love and appropriate sexual exploration as his straight peers.

It's Not Even Just About Sex

As I said at the outset of this book, many parents and advocates work tirelessly to combat the messages girls receive from the youngest ages that reduce them to their bodies or demand the illusion of perpetual sexual availability. Back when my daughter was tiny, for example, I would point to Disney heroines whose eyes were larger than their wrists ("Are *your* eyes bigger than *your* wrists? And look at her waist! Where do you think she keeps her uterus? In her purse?"). Honestly? I doubt I would've done that with a son. Yet, boys grow up in the same distorted, commodified, misogynist culture as girls; the concern over porn, while valid, can distract from the damaging impact of mainstream entertainment. Remember that, unchecked, media consumption of *any* kind is associated with greater tolerance for sexual harassment, belief in rape myths, early sexual initiation, sexual risk-taking, a greater number of partners, and stereotyping of women. Boys, too, then, need a strong counternarrative to develop grounded, realistic perspectives on women, men, sex, and love. Frankly, without it, there is a chance that they won't see women as fully human, and that they will view sex as something a female partner does *for* them and that they do *to* her. Start young, by offering little boys books, films, and other media featuring complex female protagonists. Take notice when women are absent or misrepresented on-screen or the playing field. Intervene, even if it annoys guys, to question how the media they consume presents gender roles, bodies (men's as well as women's), race, sex (is it valued or cheapened? Is there respect? Coercion? Consent?). Andrew Smiler suggests that while watching TV together, parents could periodically ask, "Would that really happen in real life?" "What is missing?" or "*Who* is missing?" That works whether characters are engaging in "romantic" behavior that could also be read as stalking, moving directly from kissing to intercourse in the span of fifteen seconds, or hooking up for the first time without any fumbling or awkwardness. One friend of mine told me she encourages her son to play his music in the car, but when she hears lyrics that are demeaning toward women or glorify violence or drugs, she turns it off and insists they talk about it. "We've never made it through more than about twenty seconds of a song," she said wryly.

Promote the Healthy and Name the Toxic

Mothers and fathers (and any other adults in a guy's life) need to challenge the unwritten rules of male socialization, the forging of masculinity through unexamined entitlement, emotional suppression, aggression, and hostility toward the feminine. Boys wouldn't stay in that "man box" if they did not reap some reward, but it is ultimately a trap: sabotaging authenticity, increasing isolation, encouraging depression, stoking rage, and promoting violence (against both others and themselves). Close relationships, whether platonic or romantic, have been found to be the number one key to personal well-being, and emotional literacy—the ability to understand and express feelings—is the key to those close relationships. Yet male conditioning renders boys numb. By adulthood, the majority of men have difficulty not only expressing but *identifying* their emotions (the formal term for that is "alexithymia"). Recall that mothers use richer emotional language with infant daughters than with their sons (and fathers an impoverished vocabulary regardless of a child's gender): that's something that isn't so hard to change. Identifying boys' emotions for them when they are small ("You seem scared." "You seem sad." "What are you feeling?" "What's going on right now?") is a start.

Fathers, by the way, are chiefs of the gender police. Boys as young as four are keenly aware of Dad's judgment, rejecting "girl" toys, even something as innocuous as miniature dishes, for fear he would think playing with them was "bad." If they are willing to stretch beyond—*way* beyond—the way they themselves were fathered, then dads (or other adult male mentors) can make a tremendous difference in sons' approach to masculinity, sex, and love. Obviously, they can lead by example (and will, whether they intend to or not), modeling the importance of affection, emotional expression, and healthy connection. But dads also play a vital role in validating boys' feelings, listening with empathy, praising a son who is himself being compassionate or caretaking. Rather than thinking about how to "toughen up" a boy (a phrase, along with "man up" and the like, that should be immediately expunged from your vocabulary), ask yourself what it might mean to raise a man who is *resilient*: who can express his feelings in a way that is respectful to himself and others as well as listen in return. To that end, among other things, fathers need to advise boys in ways beyond aggression or anger to handle conflict.

As for intimate relationships, dads can offer guidance on personal integrity; establishing and respecting sexual boundaries; mutuality; caring; pleasure. They may want to share their own evolution on some of these topics, including past mistakes and regrets. Let me reiterate: no need to be perfect, to have all the answers, or even to feel totally comfortable discussing the questions. As one college sophomore told me, "In high school, it would have made all the difference in the world to have my dad talk to me about this, even though my mom did a really good job. Because subconsciously, as a teen guy, she was still a woman telling me these things, and I really, really needed my dad to be like, 'Noah, this is real.' And because he didn't have those kinds of conversations with me, it instilled a pattern of me not having them with my friends or my partners. And I *want* to be having these conversations."

Mothers are more likely to talk with their sons about intimacy, but not as deeply or as broadly as they do with their daughters. Moms can also offer a more nuanced vision of female sexuality and explain why girls might act against their own best interests (by, say, hooking up with a guy known to be a jerk, or feeling obliged to satisfy a boy). Again, without going into excessive detail, you might consider sharing some of your own past experience, including with trauma. Women who have been sexually assaulted often feel compelled to tell their girls, even as they wrestle with when or how. But among the boys I met, hearing from a mother that she had been harmed had a life-altering impact. Obviously, that personal connection shouldn't be necessary to inspire compassion, but for teenagers, who are still relatively concrete thinkers, knowing someone in your innermost circle was affected makes an issue real (it is also a chance for moms to model resilience). A college freshman, one of just three guys in his school's anti-assault advocacy group, recalled, "When I was seventeen, my mom told me about times in college when she was drunk and men took advantage of her. I was so shocked. It was like, 'This is happening to people I care about, people I love.' That was when I realized how common it was. I will never forget that."

I would not assume, by the way, that guys know how widespread sexual misconduct, harassment, assault, and other violence against women is, at least not if adult consciousness is any measure. A nationally representative survey of 1,147 men aged eighteen to fifty-five conducted by *Glamour* and *GQ* in April 2018—a full six months after the Harvey Weinstein scandal broke—found that over half had either never heard of #MeToo or knew so little about it that they couldn't explain what it was. When, during the confirmation hearings of Supreme Court Justice Brett Kavanaugh, I asked my interview subjects how many conversations they'd had with a parent about the sexual assault allegations by Christine Blasey Ford, the typical answer was "none." Mothers *and* fathers should discuss the broad cultural impact of so-called toxic masculinity with children of all genders. The latest news stories (and there is always something) should be regular dinner table conversation, especially when they involve high school or college students. Resist reductive characterizations of "good guys" or "monsters"—those exist, but the true hard work is in the in-between. Humanize female victims: read your son the powerful impact statement by Chanel Miller, who was assaulted by Brock Turner, or the eloquent response in the *Harvard Crimson* by women who had been part of the men's soccer team's "scouting report." Talk about the precise rates of sexual harassment young women face in school hallways or on the street. Help them understand how all those comments guys make, large and small (including the "hilarious" ones), combined with popular culture and other forces, inure them to the mistreatment of women. Have your sons (and daughters) listen to the apology Dan Harmon, creator of the TV shows *Community* and *Rick and Morty*, offered on his podcast to the colleague he sexually harassed for years (she called it "a master class in how to apologize"). The point is not to shame boys but to raise awareness of how gender dynamics—mediated by race, class, sexuality, and other characteristics—function, and how they could function *much better.*

You Must, You Simply *Must*, Talk About Porn

The ubiquity of internet porn in itself means parents no longer have the luxury of squeamishness; we can no longer afford to *not* talk honestly to our children (especially our boys) about sex. The potential risk to them and to their partners is just too great. Nor do I think it's enough to dismiss porn as "not realistic" or "an adult fantasy"—that begs questions of what, exactly, is unrealistic about it and why its fantasies so frequently eroticize male aggression and female submission. Instead, as I said earlier, remind your son that curiosity about sex as well as masturbation are absolutely natural, but that porn's perspective is limited and distorted, especially for someone without much (or any) real-life context. The bodies and behaviors depicted are not typical; much of its activity would not, in truth, be pleasurable, especially for women (the ones in the videos are *paid* to pretend to enjoy what's happening and, by the way, most are left broke and unemployed within a few months). Porn can create false expectations; lower guys' satisfaction in sex and with their partners; hijack teens' sexual imaginations; eroticize gender inequality and racism. Lately, when speaking to groups of parents, mothers have asked me if I can recommend "good" porn they can offer their sons. Although you can find a variety of alternative content behind paywalls, I personally think curating your boy's porn is over the line. However, it wouldn't hurt to reiterate Emily Nagoski's take earlier in this book—the idea of masturbating in a variety of contexts, using a variety of stimuli to avoid conditioning your sexual response to porn—along with engaging in those overarching discussions about female objectification in the media and ensuring there are progressive sex-ed resources readily available in the home. In a world of fake news, media literacy, including deconstructing pornography, has value beyond sex and relationships: it is a matter of personal agency, preserving our ability to think, act, and live on our own terms.

If They Have That Solid Relationship, Consider the Sleepover

The typical response to teenage sex in the United States is "Not under my roof." But in a cross-cultural comparison of families in the US and the Netherlands, Amy Schalet found that by normalizing teen sexuality Dutch parents were able to exert *more* control over their children's behavior. That's not to say it's a free-for-all over there. Although two-thirds of Dutch teens ages fifteen to seventeen with a steady boy- or girlfriend report that the person was welcome to spend the night in their bedrooms, the Dutch actively discourage promiscuity in their children, teaching that sex should emerge from a loving relationship. Negotiating thc ground rules for those sleepovers, while admittedly cringey, provides parents another opportunity to exercise influence, reinforce values, and emphasize the need for protection. Of course, the Dutch work from an advantage: their children begin formal sex education at age four with discussions of basic body awareness—thinking about what feels good and what does not—as well as on the importance of respect for self and others in relation to family or friends. Over the years, along with curricula on anatomy, reproduction, disease prevention, contraception, and abortion, the Dutch openly address masturbation, oral sex, homosexuality, orgasm. They also stress the positive aspects of sex and relationships.

The results: even when controlling for demographic differences, Dutch teens become sexually active later than Americans, have fewer partners, are more likely to use contraception, and are less likely to say their encounters were "driven by hormones." They are more likely to say their early sexual activity took place in loving, respectful relationships in which they communicated openly with their partners about what felt good and what didn't, about what acts they wanted to engage in and what kind of protection they would need. Four out of five Dutch youth in one study said that their first sexual experiences were well timed, within their control, and fun. Eighty-six percent of girls and 93 percent of boys agreed that they and their partner "both were equally eager to have it." Compare that to the United States, where two-thirds of sexually experienced teenagers said they wished they had waited longer to have intercourse for the first time.

Decline Admission to "Dick School"

Friendship is sweet. Brotherhood is sustaining. But pay attention to the tenor of those all-male groups: under certain circumstances, they can feed aggression, antipathy toward women, and assault. When looking at colleges, inquire about options, aside from Greek life, for nighttime and weekend socializing. Boys who are thinking of joining fraternities should do their due diligence on various houses' reputations, talking not only to current members, but to unaffiliated students on campus, including girls who've attended their parties. Have student publications or local news outlets revealed incidents of sexual misconduct, racial slurs, hazing, hazardous drinking? Is sexual conquest prioritized over female dignity (a freshman at a Southern California college told me he dropped out of his frat after his pledge class was paired with a "lower-tier" sorority, so that, unbeknownst to the girls, the guys could practice their hookup technique before meeting more desirable prospects)? What programming has been put in place to educate about consent, irresponsible drinking, gender inequity, positive sexuality? How are parties made safe and comfortable for everyone, including students of color, LGBTQ+ students, and women (of any ethnicity or orientation)?

These are not trivial concerns: as I previously said, fraternity members are more likely than other boys to commit assault—as much as three times more likely, according to some research. It's unclear whether a given fraternity's disregard for women and greater tolerance for sexual misconduct is itself enough to transform otherwise nonviolent young men. What does seem to be true, though, is that high school boys who are interested in going Greek already score higher than others on proclivities for sexual aggression as well as belief in certain rape myths, such as that assault only involves physical violence or that inebriation is a reasonable excuse for male misconduct. Fraternity life validates those tendencies, makes them acceptable. That's an argument not only for ongoing, mandatory educational interventions in frat houses, but for programming geared toward boys long before college.

Athletes, too, as noted earlier, are overrepresented in sexual misconduct cases: although they make up just 4 percent of students in college, they are involved in 19 percent of assaults. "Official" players are not the only culprits; more than half of *both* intercollegiate and recreational athletes in a 2016 survey admitted to coercing a woman into sexual activity, defined as "any unwanted oral, vaginal, or anal penetration as a result of verbal or physical pressure, including rape." Teams that celebrate tribalism, aggression, and male dominance over personal integrity (as well, of course, as those that practice any form of hazing) should raise a red flag, both in terms of a young man's participation and fans' support. Athletes, who tend to hold a vaunted position in school communities, also too often know that they can act with impunity and will be protected from consequences by athletic departments, school administrations, even, potentially, by the NCAA. That must stop. Coaches, who commonly act as mentors and surrogate fathers (not to mention gatekeepers to college recruitment), are in exceptional positions of influence, especially in high schools where male athletes can dictate the social culture. They should have zero tolerance for degrading talk about women (including song lyrics blaring in the gym) or sexual braggadocio; zero tolerance for sexual misconduct; zero tolerance for calling boys (teammates or the opposition) derogatory names or teasing them about their anatomy or sexual prowess.

Guys, as I've said, will often go to great lengths to convince themselves they don't need to intervene in a troubling situation, whether it's some version of so-called locker room talk, sexual harassment, or potential assault. They rationalize. They minimize. They become passive. They convince themselves they can't actually make a difference. They laugh. They join in to avoid becoming victims themselves. Stepping up is hard and risky, especially if you are less socially powerful: strength of character can paradoxically be mocked as weak. Boys are just trying to survive, according to educator Charis Denison, to feel "safe, seen, and significant. And, if they can do that through displays of dominance and aggression, then of course that's what they're going to do." Coaches can shatter the complicity of silence, making it clear that objectifying women is *not* a masculine rite of passage. Consider a yearlong study of two thousand high school athletes that found considerably reduced rates of dating violence and support for other boys' abusive behavior among those who participated in short, weekly coach-led discussions about personal responsibility, respectful behavior, relationship abuse, insulting language, and consent. That's encouraging: all-male enclaves may be notorious as crucibles of sexism, but they could also become crucibles of change.

Over the last several years, the #MeToo movement has laid bare sexual misconduct, male privilege, and "toxic masculinity" across every sector of society. That has sparked a much-needed reckoning. It should also provide parents of boys, and boys themselves, unprecedented motivation to transform the rules of male psychological development and sexuality. That is not an easy task, but it is an exciting one: raising boys to be compassionate and egalitarian; respectful of others' boundaries; capable of connection, vulnerability, honest communication, emotional expression, and love; able to develop and sustain authentic relationships; able to be happier and more fulfilled; able to see women as true peers in the classroom, boardroom, and bedroom. Raising our boys to be the men we know they can become.

Acknowledgments

Many thanks to Jennifer Barth, Suzanne Gluck, Charis Denison, Cindy Pierce, Richard Weissbourd, Ashanti Branch, Debby Herbenick, Bryant Paul, Paul Wright, Rosalind Wiseman, Allison Stephens, Kathleen Brownbeck, Brian O'Connor, Yesenia Gorbea, Rebecca Milliman, Lara Bazelon, Brandon Nicholson, Sabina McMahon, Tony Nguyen, Deborah Roffman, Simone Marean, Rachel Simmons, D. Watkins, Aaron Retica, Laurie Abraham, Raha Nadaff, Douglas McGray, Pamela Paul, Judith Warner, Sarah Ross, Muriel Vernon, Laura Alberti, Tom Davey, Rachel Hanebutt, Rebecca Mosley, Mary Dunnewold, Virginia Witt, Thordis Elva, Leah Fessler, Himanshu Agrawa, Heather Larsen, Brandon Rein, Barbara McDonald, Tom Stranger, Peggy Edersheim Kalb, Diane Espaldon, Dan Wilson, Ina Park, Matt Dixon, Danny Sager, Brian McCarthy, Julia Sweeney, Michael Blum, Mulan Sweeney Blum, Ruth Halpern, Eva Eilenberg, Cornelia Lauf, Ayelet Waldman, Sylvia Brownrigg, Rachel Silvers, Shari Washburn, Neal Karlen, Barbara Swaiman, Leslie Orenstein, Julie Ann Orenstein, Lucy Orenstein, Debbie Orenstein, and Ari Orenstein.

To all the boys who spoke to me for this book, I owe you a

huge debt of gratitude. Whether your story is in these pages or not, I could not have written without your honesty, insight, and trust.

Special thanks to advisors Megan Ruppel, Sam Thomson, and Evelyn Wang, and to my absolutely indispensable interns: Brock Coylar, Maya Guzdar, Will Hoppin, Jacob Finkelman, and Katherine Reber.

To Steven and Daisy, all my love and boundless gratitude for your support, faith, tolerance, cheerleading, and cuisine that reigns supreme.

Notes

Chapter 1: Welcome to Dick School

11 Nearly 60 percent of American: Heilman, Barker, and Harrison, "The Man Box."

12 A 2018 national survey: Undem and Wang, "The State of Gender Equality for U.S. Adolescents."

13 Young American men also: Heilman, Barker, and Harrison, "The Man Box." See also Gallagher and Parrot, "What Accounts for Men's Hostile Attitudes Toward Women?"

13 Whether you label it: Pollack, *Real Boys*; Kimmel, *Guyland*.

13 or "the man box": Porter, "A Call to Men"; Heilman, Barker, and Harrison, "The Man Box." See also American Psychological Association, Boys and Men Guidelines Group, "Guidelines for Psychological Practice with Boys and Men."

13 Young men who most internalize: Heilman, Barker, and Harrison, "The Man Box"; June Gruber and Jessica Borelli, "The Importance of Fostering Emotional Diversity in Boys," *Scientific American*, December 12, 2017, https://tinyurl.com/y92mk2xg; Baugher and Gazmararian, "Masculine Gender Role Stress and Violence", Wong, Ringo Ho, Wang, et al., "Meta-Analyses of the Relationship Between Conformity to Masculine Norms and Mental Health-Related Outcomes"; Alia E. Dastagir, "Gender Stereotypes are Destroying Girls, and They're Killing Boys," *USA Today*, September 21, 2017, https://tinyurl.com/yae8qajo; American Psychological Association, "Sexism May Be Harmful to Men's Mental Health"; Lutz-Zois, Moler, and Brown, "Mechanisms for the Association between Traditional Masculine Ideologies and Rape Myth Acceptance Among College Men"; Gallagher and Parrot, "What Accounts for Men's Hostile Attitudes Toward Women?"

13 At its core, what psychologist: Pollack, *Real Boys*; Kimmel calls it the "Guy Code" in *Guyland.*

13 that must be continuously policed: Ibid.

15 There is no difference at birth: Weinberg, Tronick, Cohn, et al., "Gender Differences in Emotional Expressivity and Self-Regulation During Early Infancy." See also Brody, "The Socialization of Gender Differences in Emotional Expression."

15 In a classic study, adults: Condry and Condry, "Sex Differences." When told the same infant is a girl, they describe her response as "fear." Preschoolers are also more likely to attribute "disgust" to a male and "fear" to a female when read a story that is identical save the gender of the protagonist. Widen and Russell, "Gender and Preschoolers' Perception of Emotion."

15 Mothers of young children: van der Pol, Groeneveld, van Berkel, et al., "Fathers' and Mothers' Emotion Talk with Their Girls and Boys from Toddlerhood to Preschool Age"; Johnson, Caskey, Rand, et al., "Gender Differences in Adult-Infant Communication in the First Months of Life"; Aznar and Tenenbaum, "Spanish Parents' Emotion Talk and Their Children's Understanding of Emotion."

15 With sons, again, they focus: Gruber and Borelli, "The Importance of Fostering Emotional Diversity in Boys."

15 Fathers speak with less: Kristina Dell, "Mothers Talk Differently to Daughters than Sons: Study," *Time*, November 13, 2014, https://tinyurl.com/y2fcwxos; van der Pol, Groeneveld, van Berkel, et al., "Fathers' and Mothers' Emotion Talk with Their Girls and Boys from Toddlerhood to Preschool Age."

16 Despite that, according to Judy: Noah Berlatsky, "How Boys Teach Each Other to Be Boys," *The Atlantic*, June 6, 2014, https://tinyurl.com/y3rnb8mj; Vicki Zakrzewski, "Debunking the Myths About Boys and Emotions," Greatergood.com, December 1, 2014, https://tinyurl.com/y45audg3.

16 The lifelong physical and mental health consequences: Blum, Mmari, and Moreau, "It Begins at 10: How Gender Expectations Shape Early Adolescence Around the World."

16 By fourteen, boys: Pollack, *Real Boys*. See also Way, *Deep Secrets*.

16 They suspect girls: Molly Babel, Grant McGuire, and Joseph King,

"Toward a More Nuanced View of Vocal Attractiveness," *PLoS ONE*, February 19, 2014, https://tinyurl.com/y5u474ng.

16 "Emodiversity"—being able to experience: Gruber and Borelli, "The Importance of Fostering Emotional Diversity in Boys."

18 *GQ*, for instance, stipulates: Lauren Bans, "Bawl So Hard," *GQ*, June 8, 2015, https://tinyurl.com/y6alzpt4.

18 Askmen.com adds that crying: Michael A. Schottey, "The Rules for Crying in Sports," Askmen.com, https://tinyurl.com/y4228smf.

19 A survey of 150 college football: Wong, Steinfeldt, LaFollette, et al., "Men's Tears." See also Jared DeFife, "The Crying Game: Male Athletes Shedding Tears," *Psychology Today*, January 14, 2012, https://tinyurl.com/y5la566s.

21 Robert Lipsyte, a longtime: Lipsyte, "Jocks vs. Pukes," *The Nation*, July 27, 2011, https://goo.gl/eoxXfF.

21 But, Lipsyte has written: Ibid.

21 Loyalty is unconditional: Ibid. See also Steinfeldt, Foltz, Mungro, et al., "Masculinity Socialization in Sports."

23 At the same time, "fag": Pascoe, *Dude, You're a Fag*. See also Funk and Werhun, "'You're Such a Girl!'"

29 And, in fact, by the time: Paula Lavigne, "OTL: College Athletes Three Times More Likely to Be Named in Title IX Sexual Misconduct Complaints," ESPN.com, November 3, 2018, https://tinyurl.com/y5urjnbm.

30 In the spring of 2019: Shreya Chattopadhyay and Trina Paul, "2013 DU Minutes, Past Members Provide Window into Culture of Frat," *The Phoenix*, May 2, 2019, https://tinyurl.com/y689j7bl; Julie Turkewitz, "Swarthmore Fraternities Disband After Uproar Over 'Rape Attic' Documents," *New York Times*, May 1, 2019, A19; Bayliss Wagner, Naomi Park, Trina Paul, et al., "Cult of Misogyny: Leaked Internal Documents Reveal Silence Around Harmful Culture at Phi Psi," *The Phoenix*, April 18, 2019, https://tinyurl.com/y2cjzym7. For more insight into qualities of high-risk fraternities, see Boswell and Spade, "Fraternities and Collegiate Rape Culture."

32 In order for gross, crude, sexual: Jeanna Bryner, "Study Reveals Why We Laugh at Disgusting Jokes," Live Science, August 10, 2010, https://tinyurl.com/yykeeqsa.

34 Psychologist Michael Thompson has pointed out: Thompson, *Raising Cain*. See also Kimmel, *Guyland*.

Chapter 2: If It Exists, There Is Porn of It

41 It's no secret that today's children: There is even a TED Talk titled "The Great Porn Experiment."

42 One hundred million visitors: Pornhub Insights, "2018 Year in Review," Pornhub, December 11, 2018.

42 The site is owned by: David Auerbach, "Vampire Porn: MindGeek Is a Cautionary Tale of Consolidating Production and Distribution in a Single, Monopolistic Owner," Slate, October 23, 2014, https://goo.gl/5CNtjn.

43 Absent other sources: Abby Young-Powell, "Students Turn to Porn for Sex Education," *Guardian*, January 29, 2015, https://goo.gl/WMrqf8.

43 Girls in particular consult: Herbenick, Paul, Gradus, et al., The 2016 National Survey of Porn Use, Relationships, and Sexual Socialization.

43 One national survey: Ibid.

44 Often, the first exposure: See, for instance, Wolak, Mitchell, and Finkelhor, "Unwanted and Wanted Exposure to Online Pornography in a National Sample of Youth Internet Users."

44 Intentional searches started: See Sabina, Wolak, and Finkelhor, "The Nature and Dynamics of Internet Pornography Exposure for Youth."

46 That, as it happens: Dill-Shackleford, *How Fantasy Becomes Reality.*

46 Consider the college students: Ibid.

46 Karen Dill-Shackleford, a media psychologist: Ibid.

47 On the upside, there's some indication: Wright and Randall, "Pornography Consumption, Education, and Support for Same-Sex Marriage Among Adult U.S. Males." See also Regnerus, "Porn Use and Support of Same-Sex Marriage."

47 On the other hand, they're also less likely: Wright and Funk, "Pornography Consumption and Opposition to Affirmative Action for Women."

47 to disagree with or only tepidly endorse: Undem and Wang, "The State of Gender Equality for U.S. Adolescents."

47 College students who regularly: Martellozzo, Monaghan, Adler, et al., "'. . . I Wasn't Sure It Was Normal to Watch It . . .'" Other research has found that boys believe internet pornography to be more realistic than girls do and that adolescents view it as more realistic than young adults. Peter and Valkenburg, "Adolescents' Exposure to a Sexualized Media Environment and Their Notions of Women as Sex

Objects"; Peter and Valkenberg, "Adolescents' Exposure to Sexually Explicit Online Material and Recreational Attitudes Toward Sex."

47 although greater media use: O'Hara, Gibbons, Gerrard, et al., "Greater Exposure to Sexual Content in Popular Movies Predicts Earlier Sexual Debut and Increased Sexual Risk Taking"; Wright, Malamuth, and Donnerstein, "Research on Sex in the Media"; Bersamin, Bourdeau, Fisher, et al., "Television Use, Sexual Behavior, and Relationship Status at Last Oral Sex and Vaginal Intercourse"; Fisher, Hill, Grube, et al., "Televised Sexual Content and Parental Mediation"; Primack, Douglas, Fine, et al., "Exposure to Sexual Lyrics and Sexual Experience Among Urban Adolescents"; Brown, L'Engle, Pardun, et al., "Sexy Media Matter"; Zhang, Miller, and Harrison, "The Relationship Between Exposure to Sexual Music Videos and Young Adults' Sexual Attitudes"; Ward and Friedman, "Using TV as a Guide"; Ward, Hansbrough, and Walker, "Contributions of Music Video Exposure to Black Adolescents' Gender and Sexual Schemas"; Ward, "Understanding the Role of Entertainment Media in the Sexual Socialization of American Youth"; Brown, "Mass Media Influences on Sexuality"; Klein, Brown, Childers, et al., "Adolescents' Risky Behavior and Mass Media Use."

47 to have more partners: American College of Pediatricians, "The Impact of Pornography on Children," position statement, June 2016, https://goo.gl/kMokhA; Katherine Sellgren, "Pornography 'Desensitising Young People'," BBC.com, June 15, 2016, https://goo.gl/WzDeKP; Bridges, Sun, and Ezzell, "Sexual Scripts and the Sexual Behavior of Men and Women Who Use Pornography"; Sun, Johnson, and Ezzell, "Pornography and the Male Sexual Script"; Wright, "A Three-Wave Longitudinal Analysis of Preexisting Beliefs, Exposure to Pornography, and Attitude Change"; Braun-Courville and Rojas, "Exposure to Sexually Explicit Web Sites and Adolescent Sexual Attitudes and Behaviors"; Carroll, Padilla-Walker, Nelson, et al., "Generation XXX"; Peter and Valkenburg, "Adolescents' Exposure to a Sexualized Media Environment and Their Notions of Women as Sex Objects"; Maas, "The Influence of Internet Pornography on College Students"; Peter and Valkenburg, "Adolescents' Exposure to Sexually Explicit Online Material and Recreational Attitudes Toward Sex"; Bonino, Ciairano, Rabaglietti, et al., "Use of Pornography and Self-Reported Engagement in Sexual Violence Among Adolescents."

48 In 1992, 16 percent: Herbenick, Reece, Schick, et al., "Sexual Behavior in the United States." See also Susan Donaldson James, "Study Reports Anal Sex on Rise Among Teens," ABC News, December 10, 2008, https://tinyurl.com/y3cpcd53; Bahar Gholipour, "Teen Anal Sex Study," Live Science, August 14, 2014, https://goo.gl/iUiV3o.

48 most, by the way, do not: Leichliter, Chandra, Liddon, et al., "Prevalence and Correlates of Heterosexual Anal and Oral Sex in Adolescents and Adults in the United States."

48 A 2014 survey of anal sex: Marston and Lewis, "Anal Heterosex Among Young People and Implications for Health Promotion: a Qualitative Study in the UK."

49 His previous coauthored research: Those "aggressive behaviors" included hair pulling, spanking a partner hard enough to leave a mark, facial ejaculation, confinement, double penetration (penetrating a partner's anus or vagina simultaneously with another man), ass-to-mouth (anally penetrating a partner and then inserting the penis directly into her mouth), penile gagging, facial slapping, choking, and name-calling (such as "slut" or "whore"). Wright, Tokunaga, and Kraus, "Pornography Consumption and Satisfaction"; Wright, Sun, Steffen, et al., "Pornography, Alcohol, and Male Sexual Dominance."

49 Among male college students: Wright and Tokunaga, "Activating the Centerfold Syndrome"; Wright, "Show Me the Data!"

49 College students of both sexes: Brosi, Foubert, Bannon, et al., "Effects of Women's Pornography Use on Bystander Intervention in a Sexual Assault Situation and Rape Myth Acceptance"; Wright and Tokunaga, "Men's Objectifying Media Consumption, Objectification of Women, and Attitudes Supportive of Violence Against Women"; Brown and L'Engle, "X–Rated: Sexual Attitudes and Behaviors Associated with U.S. Early Adolescents' Exposure to Sexually Explicit Media"; Hald, Malamuth, and Lange, "Pornography and Sexist Attitudes Among Heterosexuals"; Foubert, Brosi, and Bannon, "Pornography Viewing Among Fraternity Men."

49 What may be of more immediate: Wright, Bridges, Sun, et al., "Personal Pornography Viewing and Sexual Satisfaction"; Wright, Steffen, and Sun, "Is the Relationship Between Pornography Consumption Frequency and Lower Sexual Satisfaction Curvilinear?"; Wright, Tokunaga, Kraus, et al., "Pornography Consumption and Satisfaction"; Wright, Sun, Steffen, et al., "Associative Pathways Between Pornography

Consumption and Reduced Sexual Satisfaction"; Morgan, "Associations Between Young Adults' Use of Sexually Explicit Materials and Their Sexual Preferences, Behaviors, and Satisfaction."

49 There is even speculation: Melissa Batchelor Warnke, "Millennials Are Having Less Sex than Any Generation in 60 years. Here's Why it Matters," *Los Angeles Times*, Aug 3, 2016, https://goo.gl/7ZpEcz; Dave Simpson, "Why Millennials Aren't Fucking," *Vice*, April 26, 2016, https://goo.gl/Nu6mvw.

57 We tend to believe: Nagoski, *Come as You Are*.

57 For men, the overlap: Ibid.

58 or some politicians: John Eligon and Michael Schwirtz, "Senate Candidate Provokes Ire With 'Legitimate Rape' Comment," *New York Times*, August 20, 2012, A13; Garance Franke-Ruta, "A Canard That Will Not Die: 'Legitimate Rape' Doesn't Cause Pregnancy," *The Atlantic*, August 19, 2012, https://tinyurl.com/y3qcm9kl.

59 Quite the opposite: Mark Shrayber, "Here's the Dangerous and Grotesque Anal Sex Trend You've Always Wanted," Jezebel, June 19, 2014, https://tinyurl.com/y3rlzkyn.

59 Even so, most heterosexual porn: Bryant Paul, author interview, December 7, 2017.

60 Pornhub's front page: Sarah Rense, "The Human Race Really Outdid Itself with Porn Searches in 2018," December 12, 2018, https://goo.gl/n9VT7r.

61 Nor is its sexual explicitness: Gamble, "From Sexual Media to Unwanted Hookups"; Ana Bridges, "Some Pornography Scenes Are Wholesome, Others Degrading," *New York Times*, November 11, 2012, https://goo.gl/NN467B; Eyal and Finnerty, "The Portrayal of Sexual Intercourse on Television"; Eyal and Kunkel, "The Effects of Sex in Television Drama Shows on Emerging Adults' Sexual Attitudes and Moral Judgments"; Kim, Sorsoli, Collins, et al., "From Sex to Sexuality"; Martino, Collins, Elliott, et al., "Exposure to Degrading Versus Nondegrading Music Lyrics and Sexual Behavior Among Youth."

62 Even a brief exposure: Davies, Spencer, Quinn, et al., "Consuming Images."

62 One clue as to how: Milburn, Mather, and Conrad, "The Effects of Viewing R-rated Movie Scenes That Objectify Women on Perceptions of Date Rape."

62 Such sexualized images: Rudman and Borgida, "The Afterglow of Construct Accessibility."

63 Female characters in: Stacy L. Smith and Crystal Allene Cook, "Gender Stereotypes: An Analysis of Popular Films and TV," The Geena Davis Institute on Gender in Media, 2008, https://tinyurl/com/jdr33g6.

63 by 2018 only 31 percent: Brent Lang, "Movies Featured More Female Protagonists in 2018, but It's Not All Good News (Study)," *Variety*, February 19, 2019, https://tinyurl.com/y6xeulgz.

63 By their teens and twenties: Horvath, Hegarty, Tyler, et al., "'Lights on at the End of the Party.'" Among college students, regular readership has also been associated with lower intent to seek sexual consent or adhere to decisions about it. Hust, Marett, Ren, et al., "Establishing and Adhering to Sexual Consent."

65 About half of boys: Undem and Wang, "The State of Gender Equality for U.S. Adolescents."

65 Nearly half also see female video game: Ibid.

66 As many as two-thirds of female journalists: Sonia Elks, "'I Will Rape You': Female Journalists Face 'Relentless' Abuse," Reuters, September 13, 2018, https://tinyurl.com/yyeavs2r.

66 Video games, incidentally: Bègue, Sarda, Gentile, et al., "Video Games Exposure and Sexism in a Representative Sample of Adolescents"; Stermer and Burkley, "SeX-Box"; Beck, Boys, Rose, et al., "Violence Against Women in Video Games"; Dill, Brown, and Collins, "Effects of Exposure to Sex-Stereotyped Video Game Characters on Tolerance of Sexual Harassment."

66 Playing sexually charged: Yao, Mahood, and Linz, "Sexual Priming, Gender Stereotyping, and Likelihood to Sexually Harass."

67 Tricia Rose, director of the Center for Study: Rose, *The Hip Hop Wars*. See also Sharpley-Whiting, *Pimps Up, Ho's Down*.

67 although hip-hop didn't invent sexism: Ibid. See also Rose, "Jay-Z—Dropping the Word 'Bitch' Doesn't Begin to Cover It"; Frank Benton, "Getting a Bad Rap: Misogynistic Themes in YouTube's Top 100 Most Viewed Pop and HipHop/Rap Music Videos by Artists' Gender," *CLA Journal* 3 (2015): 11–35; Aubrey and Frisby, "Sexual Objectification in Music Videos."

68 Dara Mathis tweeted about West: Mathis, "Dear Kanye West: North and Chicago Are Not Your Karma," Bitch Media, June 12, 2018, https://tinyurl.com/yblb52o9.

69 He blamed the weed: Brian Wu, "Does Marijuana Cause Erectile Dysfunction?" *Medical News Today*, August 9, 2018, https://tinyurl.com/y3yplxcb.

Chapter 3: Are You Experienced? Life and Love in a Hookup Culture

76 High school and college students are: Centers for Disease Control and Prevention, "Fewer U.S. High School Students Having Sex, Using Drugs," Centers for Disease Control News Room Press Release, June 14, 2018.

76 In reality, around 35 to 40 percent: Ford and England, "Hookups, Sex, and Relationships at College."

77 According to the Online College Social Life Survey: Ibid.

77 The behavior is most common: See Wade, *American Hookup*; Kuperberg and Padgett, "The Role of Culture in Explaining College Students' Selection into Hookups, Dates, and Long-Term Romantic Relationships"; Armstrong and Hamilton, *Paying for the Party*.

77 No wonder as many as 85 percent: Wade, *American Hookup*. See also Fordham News, "Sex and the Soul and the College Student," September 29, 2014, https://tinyurl.com/yxespk3n.

77 According to Lisa Wade: Wade, *American Hookup*.

78 To say that hookup culture is lubricated: Ibid.

79 Alcohol has been shown: Abbey, "Alcohol's Role in Sexual Violence Perpetration"; Davis, "The Influence of Alcohol Expectancies and Intoxication on Men's Aggressive Unprotected Sexual Intentions"; Foubert, Newberry, and Tatum, "Behavior Differences Seven Months Later"; Carr and VanDeusen, "Risk Factors for Male Sexual Aggression on College Campuses"; Abbey, Clinton-Sherrod, McAuslan, et al., "The Relationship Between the Quantity of Alcohol Consumed and the Severity of Sexual Assaults Committed by College Men"; Norris, Davis, George, et al., "Alcohol's Direct and Indirect Effects on Men's Self-Reported Sexual Aggression Likelihood"; Abbey, Zawacki, Buck, et al., "Alcohol and Sexual Assault"; Norris, George, Davis, et al., "Alcohol and Hypermasculinity as Determinants of Men's Empathic Responses to Violent Pornography."

79 drunk guys are more aggressive: Abbey, "Alcohol's Role in Sexual Violence Perpetration"; Orchowski, Berkowitz, Boggis, et al., "Bystander Intervention Among College Men."

80 That orgasm gap shrinks: Ford and England, "Hookups, Sex, and Relationships in College."

80 guys also overestimate women's orgasms: Ibid.

80 There's even some suggestion that: Bedera and Rothwell, "Emotional Intimacy and Sex"; Nicole Bedera, author interview, September 29, 2017.

81 For some college men: Wade, *American Hookup*. See also Kalish, "Masculinities and Hooking Up."

90 According to sociologist Lisa: Wade, *American Hookup*.

91 But that creates a predicament: England, Shafer, and Fogarty, "Hooking Up and Forming Romantic Relationships on Today's College Campuses"; Armstrong, England, and Fogarty, "Accounting for Women's Orgasm and Sexual Enjoyment in College Hookups and Relationships"; Armstrong, Hamilton, and England, "Is Hooking Up Bad for Young Women?"

93 In part, that was due to a dearth: In general, the gender ratio on campus is thought to influence perceptions of casual sex. Society for Personality and Social Psychology, "Imbalanced Gender Ratios Could Affect Views About Casual Sex and Hook-Up Culture," ScienceDaily, December 9, 2015, https://tinyurl.com/y5we5c76.

97 In comparing Dutch and American: Schalet, *Not Under My Roof*.

98 Yet, a large-scale survey: Giordano, Longmore, and Manning, "Gender and the Meanings of Adolescent Romantic Relationships"; Giordano, "Relationships in Adolescence."

101 romantic partners can be controlling: "Dating Abuse Statistics," Love Is Respect, https://tinyurl.com/y3n2f9wt.

102 As a result, young men: Wake Forest University, "Young Men More Vulnerable to Relationship Ups and Downs than Women," ScienceDaily, June 14, 2010, https://tinyurl.com/y2wmm3ow.

102 That said, according to sociologist: Wade, *American Hookup*.

Chapter 4: Get Used to It: Gay, Trans, and Queer Guys

107 Attitudes have shifted more: "Attitudes on Same-Sex Marriage," Pew Research Center, May 14, 2019, https://tinyurl.com/yxjjm2ro.

107 Even among groups: Ibid.

107 gay men now earn an average: Kitt Carpenter, "Gay Men Used to Earn Less than Straight Men; Now They Earn More," *Harvard Business Review*, December 4, 2017, https://tinyurl.com/y9w8cqtj.

107 As acceptance has risen: American Friends of Tel Aviv University,

"Age of 'Coming Out' Is Now Dramatically Younger," ScienceDaily, October 11, 2011, https://tinyurl.com/y4x94jxd.

107 Twenty percent of millennials: Harris Poll, "Accelerating Acceptance 2017," Los Angeles: GLAAD, 2017, https://tinyurl.com/yx8w98zu. See also Human Rights Campaign, "Growing UP LGBT," Report, Washington, DC: Human Rights Campaign, 2012, https://tinyurl.com/y69yoz5a. Twelve percent of all millenials identify as transgender or gender nonconforming, "meaning they do not identify with the sex they were assigned at birth or their gender expression is different from conventional expectations of masculinity and femininity." Of that 12 percent, 63 percent also say they do not identify as heterosexual.

107 Among Gen Z: Harris Poll, "Accelerating Acceptance 2017."

108 The 50 percent rise: "STDs in Adolescents and Young Adults," Centers for Disease Control and Prevention, https://tinyurl.com/y6yfzt9b.

108 queer boys compose one in five: Centers for Disease Control and Prevention, "HIV and Youth," September 2019, https://tinyurl.com/y5cs425y; National Center for HIV/AIDS, Viral Hepatitis, STD, and TB Prevention, "National Youth HIV and AIDS Awareness Day," Centers for Disease Control and Prevention, April 9, 2019, https://tinyurl.com/yyuzl8bx.

108 African Americans, low-income men: Ari Shapiro, "Why Men in Mississippi Are Still Dying of AIDS, Despite Existing Treatments," *All Things Considered*, National Public Radio, February 14, 2019, https://tinyurl.com/y37l6lup.

108 Abstinence-only "education": "GLSEN Calls for LGBTQ–Inclusive Sex Ed," Press Release, New York: GLSEN, 2015.

109 It also traffics in its own: Rosenberger, Reece, Schick, et al., "Sexual Behaviors and Situational Characteristics of Most Recent Male-Partnered Sexual Event among Gay and Bisexually Identified Men in the United States."

109 Although far less research: Kendall, "Educating Gay Male Youth."

112 The mainstream party culture: Wade, *American Hookup*; Lamont, Roach, and Kahn, "Navigating Campus Hookup Culture."

114 That was in 2015: Halberstam, *Trans**; Trey Taylor, "Why 2015 Was the Year of Trans Visibility," *Vogue*, December 29, 2015, https://tinyurl.com/y2xv7wfq.

114 He was not denounced or disbelieved when: Dawn Ennis, "American Medical Association Responds to 'Epidemic' of Violence Against Transgender Community,' *Forbes*, June 15, 2019, https://tinyurl.com/y4b6gxnt.

116 most transgender people: Sandy E. James, Jody Herman, Susan Rankin, et al., "The Report of the 2015 U.S. Transgender Survey," Washington, DC: National Center for Transgender Equality, 2016. Thirty-six percent of trans male respondents had undergone chest reduction surgery and 3 percent had a phalloplasty. Another 2 percent underwent metoidioplasty, which involves the creation of a phallus from the clitoris. Fourteen percent had undergone a hysterectomy. Although, according to the survey, an additional 61 percent said they would like to have chest surgery and an additional 57 percent a hysterectomy, only an additional 19 percent said they wanted phalloplasty and an additional 25 percent would like a metoidioplasty.

125 It is perhaps not surprising: Angie Leventis Lourgos, "Parents Often Struggle to Talk About Sex with LGBTQ Teens: Northwestern Study," *Chicago Tribune*, April 9, 2018, https://tinyurl.com/y3vwbvv6.

125 They need to have agency over: Ibid.

Chapter 5: Heads You Lose, Tails I Win: Boys of Color in a White World

136 Only 40 percent of African American men: Emily Tate, "Graduation Rates and Race," Inside Higher Ed, April 26, 2017, https://tinyurl.com/y9eqb6do. See also American Psychological Association, Presidential Task Force on Educational Disparities, "Ethnic and Racial Disparities in Education: Psychology's Contributions to Understanding and Reducing Disparities," Washington, DC: American Psychological Association, 2012, https://tinyurl.com/mxtg59u.

137 It turns out, according to: Armstrong and Hamilton, *Paying for the Party*.

137 In *Paying for the Party*, their five-year: Ibid.

137 Meanwhile, a sweeping: Emily Badger, Claire Cain Miller, Adam Pearce, et al., "Extensive Data Shows Punishing Reach of Racism for Black Boys," *New York Times*, March 19, 2018, https://tinyurl.com/y3m34pp7.

140 All that said, African American: Ispa-Landa, "Gender, Race, and Justifications for Group Exclusion." See also Holland, "Only Here for the Day."

141 The video plays with: Kimmel, *Guyland*. See also West, *Race Matters*.

148 South Asian women were infuriated: Aditi Natasha Kini, "I'm Tired of Watching Brown Men Fall in Love with White Women Onscreen," Jezebel, July 6, 2017, https://tinyurl.com/y3enxas3; Amil Niazi, "'The Big Sick' Is Great, and It's Also Stereotypical Toward Brown Women," *Vice*, July 7, 2017, https://tinyurl.com/yyjy9yuk.

148 Perhaps all of that sounds paranoid: "Race and Attraction, 2009–2014," OkCupid, September 9, 2014, https://tinyurl.com/ktbxmgt; *The Daily Show with Trevor Noah*, "Sexual Racism," Comedy Central, April 12, 2016, https://tinyurl.com/zt53dbc.

149 Those gender gaps play out: Gretchen Livingston and Anna Brown, "Intermarriage in the U.S. 50 Years After *Loving v. Virginia*," Pew Research Center, 2017, https://tinyurl.com/y6n6ukso.

149 Y. Joel Wong, a professor of: Wong, Horn, and Chen, "Perceived Masculinity." See also Liu and Wong, "The Intersection of Race and Gender"; Cheng, McDermott, Wong, et al., "Drive for Muscularity in Asian American Men"; Wong, Owen, Tran, et al., "Asian American Male College Students' Perceptions of People's Stereotypes About Asian American Men."

149 In interviewing a racially diverse group: Chen, "Hooked on Race." See also Shankar Vedantam, "Hookup Culture: The Unspoken Rules of Sex on College Campuses," *Hidden Brain*, National Public Radio, September 25, 2017, https://tinyurl.com/y2nzz2c6.

149 Putting aside, for a moment: See McKee, "Does Size Matter?"

153 On a college level, while: Emily Yoffe, "The Question of Race in Campus Sexual-Assault Cases," *The Atlantic*, September 11, 2017, https://tinyurl.com/y9scsam2.

154 In a 2019 nationally representative: black women and bisexual women reported the highest rates of harassment and assault. UC San Diego Center on Gender Equity and Health, Stop Street Harassment, "Measuring #MeToo: A National Study on Sexual Harassment and Assault," San Diego: UC San Diego Center on Gender Equity and Health, 2019.

155 After Bill Cosby's arrest in 2015: Feminista Jones, "Why Black Women Are Jumping to Bill Cosby's Defense—and Why They Should Stop," *Time*, January 9, 2015, https://tinyurl.com/y2s94mpn.

155 And in the final episode of: Salamishah Tillet and Scheherazade Tillet, "After the 'Surviving R. Kelly' Documentary, #MeToo Has Finally Returned to Black Girls," *New York Times,* January 10, 2019, https://tinyurl.com/ycqtfk8j.

159 They also piled up: Von Robertson and Chaney, "'I Know it [Racism] Still Exists Here'"; Adrienne Green, "The Cost of Balancing Academia and Racism," *The Atlantic,* January 21, 2016, https://tinyurl.com/y7jrwvfw; Joan Brasher, "Black College Students Face Hidden Mental Health Crisis," Research News @Vanderbilt, December 30, 2015, https://tinyurl.com/y8bpcm4f; Yasso, Smith, Ceja, et al., "Critical Race Theory, Racial Microaggressions, and Campus Racial Climate for Latina/o Undergraduates"; American Psychological Association, Presidential Task Force on Educational Disparities, "Ethnic and Racial Disparities in Education: Psychology's Contributions to Understanding and Reducing Disparities."

160 And it turns out that black: Hammond, "Taking It Like a Man."

161 At a certain point, one has to: Jake New, "Bad Apples or the Barrel?" *Inside Higher Ed,* April 15, 2015, https://tinyurl.com/y3bap7pw.

162 Additionally, while members of both: Ray, "Fraternity Life at Predominantly White Universities in the US"; Ray and Rosow, "The Two Different Worlds of Black and White Fraternity Men: Visibility and Accountability as Mechanisms of Privilege"; Ray and Rosow, "Getting Off and Getting Intimate." See also Allison and Risman, "'It Goes Hand in Hand with the Parties.'"

162 though some health educators: Zaloom, *Sex, Teens, and Everything in Between*; Charis Denison calls this "the Platinum rule." Charis Denison, author interview, June 26, 2019.

Chapter 6: I Know I'm a Good Guy, But . . .

167 "Well, something like: Zaloom, *Sex, Teens, and Everything in Between.*

169 In reality, according to: Cantor, Fisher, Chibnall, et al., *Report on the AAU Campus Climate Survey on Sexual Assault and Sexual Misconduct.*

170 In 2015, Nicole Bedera: Bedera, "Moaning and Eye Contact"; Nicole Bedera, author interview, September 29, 2017.

171 They, too, generally deny: Scully and Marolla, "Convicted Rapists' Vocabulary of Motive."

171 The truth is: Hansen, O'Byrne, and Rapley, "Young Heterosexual

Men's Use of the Miscommunication Model in Explaining Acquaintance Rape."

171 Bedera's subjects considered: Bedera, "Moaning and Eye Contact"; Nicole Bedera, author interview, September 29, 2017.

172 In 2016, researchers: Brooks, Lang-Ree, Yuan, et al., "Campus Sexual Assault." See also Becker and Tinkler, "'Me Getting Plastered and Her Provoking My Eyes'"; Jacques-Tiura, Abbey, Parkhill, et al., "Why Do Some Men Misperceive Women's Sexual Intentions More Frequently than Others Do?"; Roni Jacobson, "Targeting Drunk Women Accounts for Sexual Aggression, Not 'Blurred Lines,'" *Scientific American*, March 7, 2104, https://tinyurl.com/y2cdpr8z; Peralta, "College Alcohol Use and the Embodiment of Hegemonic Masculinity Among European American Men."

173 Women project: Maya Salam, "Women, Alcohol and Perceived 'Sexual Availability,'" *New York Times*, May 14, 2019, https://tinyurl.com/y4n2kbpk.

173 Even men who admit to: Heather Murphy, "New Scrutiny for Men Who Rape," *New York Times*, October 31, 2017, D1. See also Edwards, Bradshaw, and Hinsz, "Denying Rape but Endorsing Forceful Intercourse."

174 Other young men: Pascoe and Hollander, "Good Guys Don't Rape."

176 The narcissism of male desire: For more on how gender ideologies can complement and perpetuate one another, see Tolman, Davis, and Bowman, "'That's Just How It Is'." Zaloom calls it "the gender paradigm": Zaloom, *Sex, Teens, and Everything in Between.*

181 What's more, most: Weissbourd with Anderson, Cashin, et al., "The Talk." See also Lisa Damour, "Talking with Both Daughters and Sons About Sex," *New York Times*, January 11, 2017, https://tinyurl.com/yyoaxtju; Rothman, Miller, Terpeluk, et al., "The Proportion of U.S. Parents Who Talk with Their Adolescent Children About Dating Abuse."

182 women often freeze in response: Jim Hopper, "Freezing During Sexual Assault and Harassment," *Psychology Today*, April 3, 2018, https://tinyurl.com/y4zl8rob.

Chapter 7: All Guys Want It. Don't They?

185 In middle and high school: Reidy, Early, and Holland, "Boys Are Victims Too?"; Shaffer, Adjei, Viljoen, et al., "Ten-Year Trends in Physical Dating Violence Victimization Among Adolescent Boys and Girls in British Columbia, Canada."

185 though when it comes to the most: Kashmira Gander, "90 Percent of Teens Killed by Their Partners Are Girls—and Most of the Killers Are Men," *Newsweek*, April 15, 2019, https://tinyurl.com/y454429t.

185 When researchers ask students: Conor Friedersdorf, "The Understudied Female Sexual Predator," *The Atlantic*, November 28, 2016, https://goo.gl/4KYTsM. See also Stemple, Flores, and Meyer, "Sexual Victimization Perpetrated by Women"; French, Tilghman, and Malebranche, "Sexual Coercion Context and Psychosocial Correlates Among Diverse Males"; Fisher and Pina, "An Overview of the Literature on Female-Perpetrated Adult Male Sexual Victimization"; Turchik, "Sexual Victimization Among Male College Students"; Kaestle, "Sexual Insistence and Disliked Sexual Activities in Young Adulthood"; Stemple, "Male Rape and Human Rights."

185 A three-year, $2.2 million: Mellins, Walsh, Sarvet, et al., "Sexual Assault Incidents among College Undergraduates."

185 Other research has found: John Foubert, "'Rapebait' E-mail Reveals Dark Side of Frat Culture," CNN, October 9, 2013, https://tinyurl.com/yxz7jecj.

189 They, too, would rather feel violated: Ford, "Describing Unwanted Sex with Women"; Ford, "'Going with the Flow'"; Ford and Soto-Marquez, "Sexual Assault Victimization Among Straight, Gay/Lesbian, and Bisexual College Students."

189 In one study of college students: Muehlenhard and Shippee, "Men's and Women's Reports of Pretending Orgasm." See also Roberts, Kippax, and Waldby, "Faking It."

190 Their accounts also typically: Ford, "'Going with the Flow'."

191 When news broke of a Florida: Laurie Roberts, "Boy Allegedly Molested by Goodyear Teacher Brittany Zamora Is . . . Lucky?" AZCentral, March 26, 2018, https://tinyurl.com/y6r7kj3v; Stewart Perrie, "Why Is There a Double Standard When Female Teachers Have Sex with Students?" Lad Bible, March 6, 2018, https://tinyurl.com/y4727wl8; Hollie McKay, "Female Teachers Having Sex with Students: Double Standards, Lack of Awareness," *Fox News*, June 30, 2017, https://tinyurl.com/y55uhj67.

191 Shortly before he won: Marlow Stern, "'The Daily Show' Digs Up Creepy Clip of Trump Defending a Statutory Rape," The Daily Beast, September 29, 2016, https://tinyurl.com/y5uesyap.

192 as many as one in six boys: Dube, Anda, Whitfield, et al., "Long-term Consequences of Childhood Sexual Abuse by Gender of Victim."

192 A freshman at Brown: Emily Kassie, "Male Victims of Campus Sexual Assault Speak Out 'We're Up Against a System That's Not Designed to Help Us,'" *Huffington Post,* January 27, 2105, https://tinyurl.com/yxkc6c8v.

193 The SHIFT study found that: Khan, Hirsch, Wamboldt, et al., "'I Didn't Want to Be "That Girl"': The Social Risks of Labeling, Telling, and Reporting Sexual Assault."

Chapter 8: A Better Man

198 A national survey of students: Cantor, Fisher, Chibnall, et al., *Report on the AAU Campus Climate Survey on Sexual Assault and Sexual Misconduct.* See also Carey, Durney, Shepardson, et al., "Incapacitated and Forcible Rape of College Women."

198 One woman, whose son: Anemona Hartocollis and Christina Capecchi, "Mothers 'Willing to Do Everything,' Mothers Defend Sons Accused of Sexual Assault," *New York Times,* October 24, 2017, A12.

199 Another, who was among a: Ibid.

212 Although its primary emphasis: For more on best practices in restorative justice, see Karp, *The Little Book of Restorative Justice for Colleges and Universities*; Bargen, Edwards, Hartman, et al., *Serving Crime Victims Through Restorative Justice*; Kaplan, "Restorative Justice and Campus Sexual Misconduct."

212 In his study of 659: Karp and Sacks, "Student Conduct, Restorative Justice, and Student Development."

212 What's more, as Judith: Herman, "Justice from the Victim's Perspective."

212 There is no perfect system: See, for instance, Yung, "Concealing Campus Sexual Assault."

Chapter 9: Deep Breath: Talking to Boys

219 Yet, realistically, only twenty-four states: Christina Capatides, "A Cup Full of Spit, a Chewed Up Piece of Gum. These Are the Metaphors Used to Teach Kids About Sex," CBS News, April 29, 2019, https://tinyurl.com/y5cg744y.

219 Well, right this minute: Ibid. See also, US House of Representatives, *The Content of Federally Funded Abstinence-Only Education Programs*; Santelli, Kantor, Grilo, et al., "Abstinence-Only-Until-Marriage."

220 despite a federal investment in abstinence-only: Jessica Boyer, "New Name, Same Harm," Guttmacher Policy Review, February 28, 2018, https://tinyurl.com/y5evxj3h; Andrea Zelinski, "Rewrite of Texas Sex Education Standards Could Include Lessons on Contraception, Gender Identity," *Houston Chronicle*, June 13, 2019, https://tinyurl.com/y4nekl8f; Advocates for Youth, "Sexual Education: Research and Results," Fact Sheet, Washington, DC: Advocates for Youth, 2009, https://tinyurl.com/y5uf7g97.

220 Equally concerning, while pleasure-based: Santelli, Grilo, Choo, et al., "Does Sex Education Before College Protect Students from Sexual Assault in College?"; Tina Rosenberg, "Equipping Women to Stop Campus Rape," *New York Times*, May 30, 2018, https://tinyurl.com/y4brgua7.

220 Fewer than half of high schools: Planned Parenthood, "What's the State of Sex Education in the U.S.?" https://tinyurl.com/yblzzxpa; Centers for Disease Control and Prevention, "16 Critical Sexual Education Topics," https://tinyurl.com/y2jve6kd.

220 only ten states require that: Laura Fay, "Just 24 States Mandate Sex Education for K-12 Students, and Only 9 Require any Discussion of Consent. See How California Compares," LA School Report, April 1, 2019, https://tinyurl.com/yxvhey73. Since the publication of this article, Colorado also began mandating some form of consent education when sex education is taught in the state's schools.

221 decades of research have: Widman, Evans, and Javidi, "Assessment of Parent-Based Interventions for Adolescent Sexual Health: A Systematic Review and Meta-Analysis.

221 Despite their eye-rolling: Holman and Koenig Kellas, "'Say Something Instead of Nothing'"; Weissbourd with Anderson, Cashin, et al., "The Talk"; Roni Caryn Rabin, "Why Parents Should Have the 'Sex Talk' with Their Children," *New York Times*, November 4, 2015, https://tinyurl.com/y3e2pomf; Alexandra Ossola, "Kids Really Do Want to Have 'The Talk' with Parents," *Popular Science*, March 5, 2015, https://tinyurl.com/yx8pyham.

224 As health educator Shafia Zaloom: Shafia Zaloom, *Sex, Teens, and Everything in Between.*

224 Along with sexual information: Weissbourd with Anderson, Cashin, et al., "The Talk"; Rick Weissbourd, "Teens' Romantic Relationships,"

Making Caring Common Project, October 2018, https://tinyurl.com/yyllgkdt; Ossola, "Kids Really Do Want to Have 'The Talk' with Parents."

225 In contrast to the "always down for it": Andrew Smiler, "Why Do Boys Date and Have Sex?" Andrewsmiler.com, January 23, 2015, https://tinyurl.com/y4wo7vad.

226 Amy Schalet, a sociologist who: Amy Schalet, "The New ABCD's of Talking About Sex with Teenagers," HuffPost, November 2, 2011, https://tinyurl.com/y5sdodh5; Orenstein, *Girls & Sex*.

227 Andrew Smiler suggests: Smiler, *Challenging Casanova*.

228 Close relationships, whether platonic or romantic: Jody Aked and Sam Thompson, "Five Ways to Well-being: New Applications, New Ways of Thinking," New Economics Foundation, July 5, 2011, https://tinyurl.com/y2rrc89g.

228 By adulthood, the majority of men: Levant, Allen, and Lien, "Alexithymia in Men"; Karakis and Levant, "Is Normative Male Alexithymia Associated with Relationship Satisfaction, Fear of Intimacy and Communication Quality Among Men in Relationships?" See also "How to Encourage Healthy Emotional Development in Boys," https://tinyurl.com/yyzg9r6c.

228 Recall that mothers: Gruber and Borelli, "The Importance of Fostering Emotional Diversity in Boys."

229 Boys as young as four are keenly aware: Cherney and Dempsey, "Young Children's Classification, Stereotyping and Play Behaviour for Gender Neutral and Ambiguous Toys."

231 A nationally representative: Editors of *GQ*, "What 1,147 Men Think About #MeToo: A *Glamour* x *GQ* Survey," *GQ*, May 30, 2018, https://tinyurl.com/y64r3db9.

231 read your son the powerful: Katie J. M. Baker, "Here's the Powerful Letter the Stanford Victim Read to Her Attacker," BuzzFeed News, June 3, 2016, https://tinyurl.com/yd5xdase.

231 or the eloquent response: Kelsey Clayman, Brooke Dickens, Alika Keene, et al., "Stronger Together," *Harvard Crimson*, October 29, 2016, https://tinyurl.com/y4cn842k.

231 Have your sons (and daughters): Dan Harmon, "Don't Let Him Wipe or Flush," *Harmontown*. The apology starts just past the nineteen minute mark. See also Nancy Updike, "Finally," *This American Life*, National Public Radio, May 10, 2019, https://tinyurl.com/y5jjla96.

232 the ones in the videos are *paid*: Jill Bauer and Ronna Gradus, *Hot Girls Wanted*, Netflix, 2015.

233 But in a cross-cultural comparison: Schalet, *Not Under My Roof.*

233 their children begin formal: Ibid. See also Orenstein, *Girls & Sex*; Saskia de Melker, "The Case for Starting Sex Education in Kindergarten," PBS News Hour, May 27, 2015, https://tinyurl.com/ycjsfx5o.

234 The results: even when: Brugman, Caron, and Rademakers, "Emerging Adolescent Sexuality"; Schalet, *Not Under My Roof.*

234 Four out of five: Brugman, Caron, and Rademakers, "Emerging Adolescent Sexuality."

234 Compare that to: Bill Albert, With One Voice 2012: America's Adults and Teens Sound Off About Teen Pregnancy, Washington, DC: The National Campaign to Prevent Teen and Unplanned Pregnancy, 2012, https://tinyurl.com/y4bnjvyg.

234 But pay attention to the tenor of: Sanday, "Rape-Prone Versus Rape-Free Campus Cultures."

234 Boys who are thinking of joining: Two essential reads: Caitlin Flanagan, "The Dark Power of Fraternities," *The Atlantic*, March 2014, https://tinyurl.com/yatzxnhe; Armstrong and Sweeney, "Sexual Assault on Campus."

235 These are not trivial concerns: John Foubert, "'Rapebait' E-mail Reveals Dark Side of Frat Culture," CNN, October 9, 2013, https://tinyurl.com/yxz7jecj. See also Seabrook, Ward, and Giaccardi, "Why Is Fraternity Membership Associated with Sexual Assault?"; Martinez, Wiersma-Mosley, Jozkowski, et al., "'Good Guys Don't Rape'"; Seabrook and Ward, "Bros Will Be Bros"; Mellins, Walsh, Sarvet, et al., "Sexual Assault Incidents Among College Undergraduates."

235 What does seem to be true, though: Seabrook, Ward, and Giaccardi, "Why Is Fraternity Membership Associated with Sexual Assault?"; Seabrook and Ward, "Bros Will Be Bros?"

235 Athletes, too, as noted earlier: Paula Lavigne, "OTL: College Athletes Three Times More Likely to Be Named in Title IX Sexual Misconduct Complaints"; Lisa Wade, "Rape on Campus: Athletes, Status, and the Sexual Assault Crisis," *The Conversation*, March 6, 2017, https://tinyurl.com/y469652p.

235 "Official" players are not: Jake New, "Sexual Coercion Among Athletes," Inside Higher Ed, June 3, 2016, https://tinyurl.com/hcffen3.

236 Teams that celebrate: See Sanday, "Rape-Prone Versus Rape-Free Campus Cultures."

236 Athletes, who tend to hold: Lisa Wade, "Rape on Campus"; Jake New, "The 'Black Hole' of College Sports," Inside Higher Ed, February 9, 2017, https://tinyurl.com/y2mklqp5; Kareem Abdul-Jabbar, "Colleges Need to Stop Protecting Sexual Predators," January 31, 2015, *Time,* https://tinyurl.com/y59m3sj3; "College Athletic Departments' Role in Investigating Sexual Assaults," Athletic Business, April 2015, https://tinyurl.com/y2nv367h. See also Nancy Armour, "Opinion: NCAA Continues to Drop the Ball by Accepting Athletes Punished for Sexual Assault," *USAToday,* April 4, 2019, https://tinyurl.com/yyljkt5q; Bernard Lefkowitz, *Our Guys,* New York: Vintage, 1998.

237 Consider a yearlong study: Miller, Tancredi, McCauley, et al., "'Coaching Boys into Men'." See also Futures Without Violence, "Coaching Boys into Men," https://tinyurl.com/y4nsrpzc.

Bibliography

Abbey, Antonia. "Alcohol's Role in Sexual Violence Perpetration: Theoretical Explanations, Existing Evidence, and Future Directions." *Drug and Alcohol Review* 30, no. 5 (2011): 481–89.

Abbey, Antonia, A. Monique Clinton-Sherrod, Pam McAuslan, et al. "The Relationship Between the Quantity of Alcohol Consumed and the Severity of Sexual Assaults Committed by College Men." *Journal of Interpersonal Violence* 18, no. 7 (2003): 813–33.

Abbey, Antonia, Tina Zawacki, Philip O. Buck, et al. "Alcohol and Sexual Assault." *Alcohol Research and Health* 25, no. 1 (2001): 43–51.

Allison, Rachel, and Barbara J. Risman. "A Double Standard for 'Hooking Up': How Far Have We Come Toward Gender Equality?" *Social Science Research* 42, no. 5 (2013): 1191–206.

———. "'It Goes Hand in Hand with the Parties': Race, Class, and Residence in College Student Negotiations of Hooking Up." *Sociological Perspectives* 57, no. 1 (2014): 102–23.

American Psychological Association, Boys and Men Guidelines Group. "Guidelines for Psychological Practice with Boys and Men." Washington, DC: American Psychological Association, 2018.

Armstrong, Elizabeth A., and Laura T. Hamilton. *Paying for the Party: How College Maintains Inequality.* Cambridge, MA: Harvard University Press, 2015.

Armstrong, Elizabeth A., Paula England, and Alison C. K. Fogarty. "Accounting for Women's Orgasm and Sexual Enjoyment in College Hookups and Relationships." *American Sociological Review* 77 (2012): 435–62.

Armstrong, Elizabeth A., Laura T. Hamilton, and Paula England. "Is Hooking Up Bad for Young Women?" *Contexts* 9, no. 3 (2010): 22–27.

Armstrong, Elizabeth, and Brian Sweeney. "Sexual Assault on Campus: A Multilevel, Integrative Approach to Party Rape." *Social Problems* 53, no. 4 (2006): 483–99.

Aubrey, Jennifer Stevens, and Cynthia Frisby. "Sexual Objectification in Music Videos: A Content Analysis Comparing Gender and Genre." *Mass Communication and Society* 14 (2011): 475–501.

Bargen, Catherine, Alan Edwards, Matthew Hartman, et al. *Serving Crime Victims Through Restorative Justice A Resource Guide for Leaders and Practitioners.* Alberta, Canada: Alberta Restorative Justice Association, 2018.

Baugher, Amy, and Julie A. Gazmararian. "Masculine Gender Role Stress and Violence: A Literature Review and Future Directions." *Aggression and Violent Behavior* 24 (2015): 107–112.

Beck, Victoria Simpson, Stephanie Boys, Christopher Rose, et al. "Violence Against Women in Video Games: A Prequel or Sequel to Rape Myth Acceptance?" *Journal of Interpersonal Violence* 27, no. 15 (2012): 3016–31.

Becker, Sarah, and Justine Tinkler. "'Me Getting Plastered and Her Provoking My Eyes': Young People's Attribution of Blame for Sexual Aggression in Public Drinking Spaces." *Feminist Criminology* 10, no. 3 (2015): 235–58.

Bedera, Nicole. "Moaning and Eye Contact: College Men's Negotiations of Sexual Consent in Theory and in Practice." SocArXiv, August 12, 2017. https://tinyurl.com/ybwht2hl.

Bedera, Nicole, and William Rothwell. "Emotional Intimacy and Sex: Exploring Sources of Men's Sexual Pleasure in Hookups and Romantic Relationships." Montreal: American Sociological Association Annual Conference, August 2017.

Bègue, Laurent, Elisa Sarda, Douglas A. Gentile, et al. "Video Games Exposure and Sexism in a Representative Sample of Adolescents." *Frontiers in Psychology* 8 (2017): 466. https://goo.gl/KFZsWd.

Bersamin, Melina M., Beth Bourdeau, Deborah A. Fisher, et al. "Television Use, Sexual Behavior, and Relationship Status at Last Oral Sex and Vaginal Intercourse." *Sexuality and Culture* 14 (2010): 157–68.

Blum, Robert W., Kristin Mmari, and Caroline Moreau. "It Begins at 10: How Gender Expectations Shape Early Adolescence Around the World." *Journal of Adolescent Health* 61, no. 4 (2017): S3–S4.

Bonino, Silvia, Silvia Ciairano, Emanuela Rabaglietti, et al. "Use of Pornography and Self-Reported Engagement in Sexual Violence Among Adolescents." *European Journal of Developmental Psychology* 3 (2006): 265–88.

Boswell, A. Ayres, and Joan Z. Spade. "Fraternities and Collegiate Rape Culture: Why Are Some Fraternities More Dangerous Places for Women?" *Gender and Society* 10, no. 2 (1996): 133–47.

Braun-Courville, Debra, and Mary Rojas. "Exposure to Sexually Explicit Web Sites and Adolescent Sexual Attitudes and Behaviors." *Journal of Adolescent Health* 45, no. 2 (2009): 156–62.

Brody, Leslie R. "The Socialization of Gender Differences in Emotional Expression: Display Rules, Infant Temperament, and Differentiation." In *Gender and Emotion: Social Psychological Perspectives,* edited by Agneta Fischer. Cambridge: Cambridge University Press, 2000, 24–47.

Bridges, Ana, Chyng F. Sun, Matthew B. Ezzell, et al. "Sexual Scripts and the Sexual Behavior of Men and Women Who Use Pornography." *Sexualization, Media, & Society* 2, no. 4 (2016). https://doi.org/10.1177/2374623816668275.

Brooks, Tess, Cecilia Lang-Ree, Helen Yuan, et al. "Campus Sexual Assault: Conflicting Expectations and Beliefs." Cambridge, MA: Confi, 2016. https://goo.gl/i1do3W.

Brosi, Matthew, John D. Foubert, R. Sean Bannon, et al. "Effects of Women's Pornography Use on Bystander Intervention in a Sexual Assault Situation and Rape Myth Acceptance." *Oracle: The Research Journal of the Association of Fraternity/Sorority Advisors* 6, no. 2 (2011): 26–35.

Brown, Jane D. "Mass Media Influences on Sexuality." *Journal of Sex Research* 39, no. 1 (2002): 42–45.

Brown, Jane D., and Kelly L. L'Engle. "X-Rated: Sexual Attitudes and Behaviors Associated with U.S. Early Adolescents' Exposure to Sexually Explicit Media." *Communication Research* 36, no. 1 (2009): 129–51.

Brown, Jane D., Kelly L. L'Engle, Carol Pardun, et al. "Sexy Media Matter: Exposure to Sexual Content in Music, Movies, Television, and Magazines Predicts Black and White Adolescents' Sexual Behavior." *Pediatrics* 117, no. 4 (2006): 1018–27.

Brugman, Margaret, Sandra L. Caron, and Jany Rademakers. "Emerging Adolescent Sexuality: A Comparison of American and Dutch College Women's Experiences." *International Journal of Sexual Health* 22, no. 1 (2010): 32–46.

Cantor, David, Bonnie Fisher, Susan Chibnall, et al. *Report on the AAU Campus Climate Survey on Sexual Assault and Sexual Misconduct.* Washington, DC: Association of American Universities, 2015. https://goo.gl/hiXGI8.

Carey, Kate, Sarah Durney, Robyn Shepardson, et al. "Incapacitated and Forcible Rape of College Women: Prevalence Across the First Year." *Journal of Adolescent Health* 56 (2015): 678–80.

Carr, Joetta L., and Karen M. VanDeusen. "Risk Factors for Male Sexual Aggression on College Campuses." *Journal of Family Violence* 19, no. 5 (2004): 279–89.

Carroll, Jason S., Laura M. Padilla-Walker, Larry J. Nelson, et al. "Generation XXX: Pornography Acceptance and Use Among Emerging Adults." *Journal of Adolescent Research* 23 (2008): 6–30.

Chen, Nicole. "Hooked on Race: An Investigation of the Racialized Hookup Experiences of White, Asian, and Black College Women." Unpublished manuscript, University of Michigan, 2014.

Cheng, Hsiu-Lan, Ryon C. McDermott, Y. Joel Wong, et al. "Drive for Muscularity in Asian American Men: Sociocultural and Racial/Ethnic Factors as Correlates." *Psychology of Men & Masculinity* 17, no. 3 (2016): 215–27.

Cherney, Isabelle D., and Jessica Dempsey. "Young Children's Classification, Stereotyping and Play Behaviour for Gender Neutral and Ambiguous Toys." *Educational Psychology* 30, no. 6 (2010): 651–69.

Condry, John, and Sandra Condry. "Sex Differences: A Study of the Eye of the Beholder." *Child Development* 47, no. 3 (1976): 812–19.

Cooper, Al. "Sexuality and the Internet: Surfing into the New Millennium." *CyberPsychology & Behavior* 1, no. 2 (2009).

Davies, Paul, Steven J. Spencer, Diane M. Quinn, et al. "Consuming Images: How Television Commercials That Elicit Stereotype Threat Can Restrain Women Academically and Professionally." *Personality and Social Psychology Bulletin* 28, no. 12 (2002): 1615–28.

Davis, Kelly Cue. "The Influence of Alcohol Expectancies and Intoxication on Men's Aggressive Unprotected Sexual Intentions." *Experimental and Clinical Psychopharmacology* 18, no. 5 (2010): 418–28.

Dill, Karen E., Brian P. Brown, and Michael A. Collins. "Effects of Exposure to Sex-Stereotyped Video Game Characters on Tolerance of Sexual Harassment." *Journal of Experimental Social Psychology* 44, no. 5 (2008): 1402–08.

Dill-Shackleford, Karen E. *How Fantasy Becomes Reality: Information and Entertainment Media in Everyday Life, Revised and Expanded.* New York: Oxford University Press, 2016.

Dube, Shanta, Robert Anda, Charles Whitfield, et al. "Long-Term Consequences of Childhood Sexual Abuse by Gender of Victim." *American Journal of Preventive Medicine* 28 (2005): 430–38.

Edwards, Sarah R., Kathryn A. Bradshaw, and Verlin B. Hinsz. "Denying Rape but Endorsing Forceful Intercourse: Exploring Differences Among Responders." *Violence and Gender* 1, no. 4 (2014): 188–93.

England, Paula, Emily Fitzgibbons Shafer, and Alison C. K. Fogarty. "Hooking Up and Forming Romantic Relationships on Today's College Campuses." In *Gendered Society Reader,* 3rd ed., edited by Michael S. Kimmel and Amy Aronson. New York: Oxford University Press, 2008.

Eyal, Keren, and Keli Finnerty. "The Portrayal of Sexual Intercourse on Television: How, Who, and with What Consequence?" *Mass Communication and Society* 12, no. 2 (2009).

Eyal, Keren, and Dale Kunkel. "The Effects of Sex in Television Drama Shows on Emerging Adults' Sexual Attitudes and Moral Judgments." *Journal of Broadcasting & Electronic Media* 52, no. 2 (2008): 161–81.

Fisher, Deborah A., Douglas L. Hill, Joel W. Grube, et al. "Televised Sexual Content and Parental Mediation: Influences on Adolescent Sexuality." *Media Psychology* 12, no. 2 (2009): 121–47.

Fisher, Nicola L., and Afroditi Pina. "An Overview of the Literature on Female-Perpetrated Adult Male Sexual Victimization." *Aggression and Violent Behavior* 18, no. 1 (2013).

Ford, Jessie. "Describing Unwanted Sex with Women: How Heterosexual Men's Accounts Uphold Masculinity." Unpublished paper, 2018.

———. "'Going with the Flow': How College Men's Experiences of Unwanted Sex Are Produced by Gendered Interactional Pressures." *Social Forces* 96, no. 3 (2018): 1303–24.

Ford, Jessie, and Paula England. "Hookups, Sex, and Relationships at College." Contexts.com, December 22, 2014. https://tinyurl.com/y67vmmqv.

Ford, Jessie, and José G. Soto-Marquez. "Sexual Assault Victimization Among Straight, Gay/Lesbian, and Bisexual College Students." *Violence and Gender* 3, no. 2 (2016): 107–15.

Fortenberry, Dennis J., Vanessa Schick, Debby Herbenick, et al. "Sexual Behaviors and Condom Use at Last Vaginal Intercourse: A National Sample of Adolescents Ages 14 to 17 Years." *Journal of Sexual Medicine* 7, supplement 5 (2010): 305–14.

Foubert, John D., Matthew W. Brosi, and R. Sean Bannon. "Pornography Viewing Among Fraternity Men: Effects on Bystander Intervention, Rape Myth Acceptance and Behavioral Intent to Commit Sexual Assault." *Sexual Addiction and Compulsivity: The Journal of Treatment and Prevention* 18, no. 4 (2011): 212–31.

Foubert, John D., Jonathan T. Newberry, and Jerry L. Tatum. "Behavior Differences Seven Months Later: Effects of a Rape Prevention Program on First-Year Men Who Join Fraternities." *NASPA Journal* 44 (2007): 728–49.

French, Bryana H., Jasmine D. Tilghman, and Dominique A. Malebranche. "Sexual Coercion Context and Psychosocial Correlates Among Diverse Males." *Psychology of Men & Masculinity* 16, no. 1 (2015): 42–53.

Funk, Leah C., and Cherie D. Werhun. "'You're Such a Girl!' The Psychological Drain of the Gender-Role Harassment of Men." *Sex Roles* 65, no. 1–2 (2011).

Gallagher, Kathryn E., and Dominic J. Parrot. "What Accounts for Men's Hostile Attitudes Toward Women? The Influence of Hegemonic Male Role Norms and Masculine Gender Role Stress." *Violence Against Women* 17, no. 5 (2011): 568–83.

Gamble, Hillary. "From Sexual Media to Unwanted Hookups: The Mediating Influence of College Students' Endorsement of Traditional Heterosexual Scripts, Sexual Self-Concept, and Perceived Peer Norms." Unpublished doctoral dissertation. Tucson, AZ: University of Arizona, 2016.

Giordano, Peggy C, Monica A. Longmore, and Wendy D. Manning. "Gender and the Meanings of Adolescent Romantic Relationships: A Focus on Boys." *American Sociological Review* 71 (2006): 260–87.

———. "Relationships in Adolescence." *Annual Review of Sociology* 29 (2003): 257–81.

Halberstam, Jack. *Trans**: A Quick and Quirky Account of Gender Variability.* Berkeley: University of California Press, 2018.

Hald, Gert Martin, Neil N. Malamuth, and Theis Lange. "Pornography and Sexist Attitudes Among Heterosexuals." *Journal of Communication* 63, no. 4 (2013): 638–60.

Hammond, Wisdom Powell. "Taking It Like a Man: Masculine Role Norms as Moderators of the Racial Discrimination-Depressive Symptoms Association Among African American Men." *American Journal of Public Health* 102, supplement 2 (2012): S232–41.

Hansen, Susan, Rachael O'Byrne, and Mark Rapley. "Young Heterosexual Men's Use of the Miscommunication Model in Explaining Acquaintance Rape." *Sexuality Research and Social Policy* 7 (2010): 45–49.

Heilman, Brian, Gary Barker, and Alexander Harrison. *The Man Box: A Study on Being a Young Man in the US, UK, and Mexico.* Washington, DC, and London: Promundo-US and Unilever, 2017.

Herbenick, Debby, Bryant Paul, Ronna Gradus, et al. The 2016 National Survey of Porn Use, Relationships, and Sexual Socialization. Indiana University.

Herbenick, Debby, Michael Reece, Vanessa Schick, et al. "Sexual Behavior in the United States: Results from a National Probability Sample of Men and Women Ages 14–94." *Journal of Sexual Medicine* 7, supplement 5 (2010): 255–65.

Herman, Judith Lewis. "Justice from the Victim's Perspective." *Violence Against Women* 11, no. 5 (2005): 571–602.

Hirsch, Jennifer S., Shamus R. Khan, Alexander Wamboldt, et al. "Social Dimensions of Sexual Consent Among Cisgender Heterosexual College Students: Insights from Ethnographic Research." *Journal of Adolescent Health* 64, no. 1 (2019): 26–35.

Holland, Megan M. "Only Here for the Day: The Social Integration of Minority Students at a Majority White High School." *Sociology of Education* 85, no. 2 (2012): 101–20.

Holman, Amanda, and Jody Koenig Kellas. "'Say Something Instead of Nothing': Adolescents' Perceptions of Memorable Conversations About Sex-Related Topics with Their Parents." *Communication Monographs* 85, no. 3 (2018): 357–79.

Horvath, Miranda H., Peter Hegarty, Suzannah Tyler, et al. "'Lights on at the End of the Party': Are Lads' Mags Mainstreaming Dangerous Sexism?" *British Journal of Psychology* 103, no. 4 (2012).

Hust, Stacey J. T., Emily Garrigues Marett, Chunbo Ren, et al. "Establishing and Adhering to Sexual Consent: The Association between Reading Magazines and College Students' Sexual Consent Negotiation." *Journal of Sex Research* 51, no. 3 (2014): 280–90.

Ispa-Landa, Simone. "Gender, Race, and Justifications for Group Exclusion: Urban Black Students Bussed to Affluent Suburban Schools," *Sociology of Education* 86, no. 3 (2013): 218–33.

Jacques-Tiura, Angela, Antonia Abbey, Michele R. Parkhill, et al. "Why Do Some Men Misperceive Women's Sexual Intentions More Frequently Than Others Do? An Application of the Confluence Model." *Personality and Social Psychology Bulletin* 33, no. 11 (2007): 1467–80. https://goo.gl/D8vMjx.

Kaestle, Christine Elizabeth. "Sexual Insistence and Disliked Sexual Activities in Young Adulthood: Differences by Gender and Relationship Characteristics." *Perspectives on Sexual and Reproductive Health* 41, no. 1 (2009): 33–39.

Kalish, Rachel. "Masculinities and Hooking Up: Sexual Decision-Making at College." *Culture, Society and Masculinities* 5, no. 2 (2013): 147–65.

Kaplan, Margo. "Restorative Justice and Campus Sexual Misconduct." *Temple Law Review* 89, no. 4 (2017): 701–45.

Karakis, Emily N., and Ronald F. Levant. "Is Normative Male Alexithymia Associated with Relationship Satisfaction, Fear of Intimacy and Communication Quality Among Men in Relationships?" *Journal of Men's Studies* 20, no. 3 (2012): 179–86.

Karp, David R. *The Little Book of Restorative Justice for Colleges and Universities: Repairing Harm and Rebuilding Trust in Response to Student Misconduct*. New York: Good Books, 2015.

Karp, David R., and Casey Sacks. "Student Conduct, Restorative Justice, and Student Development: Findings from the STARR Project: A Student Accountability and Restorative Research Project." *Contemporary Justice Review* 17, no. 2 (2014): 154–72.

Kendall, Christopher N. "Educating Gay Male Youth: Since When Is Pornography a Path Self-Respect?" *Journal of Homosexuality* 47, no. 3–4 (2004): 83–128.

Khan, Shamus R., Jennifer S. Hirsch, Alexander Wamboldt, et al. "'I Didn't Want to Be "That Girl"': The Social Risks of Labeling, Telling, and Reporting Sexual Assault." *Sociological Science* 5, no. 19. https://tinyurl.com/y2klpw5o.

Kim, Janna, C. Lynn Sorsoli, Katherine Anne Collins, et al. "From Sex to Sexuality: Exposing the Heterosexual Script on Primetime Network Television." *Journal of Sex Research* 44, no. 2 (2007).

Kimmel, Michael. *Guyland: The Perilous World in Which Boys Become Men.* New York: Harper Perennial, 2009.

Klein, Jonathan D., Jane Brown, Kim Childers, et al. "Adolescents' Risky Behavior and Mass Media Use." *Pediatrics* 92, no. 1 (1993): 24–31.

Kuperberg, Arielle, and Joseph E. Padgett. "The Role of Culture in Explaining College Students' Selection into Hookups, Dates, and Long-Term Romantic Relationships." *Journal of Social and Personal Relationships* 33, no. 8 (2016): 1070–96.

Lamont, Ellen, Teresa Roach, and Sope Kahn. "Navigating Campus Hookup Culture: LGBTQ Students and College Hookups." *Sociological Forum* 33, no. 4 (2018): 1000–22.

Leichliter, Jami, Anjani Chandra, Nicole Liddon, et al. "Prevalence and Correlates of Heterosexual Anal and Oral Sex in Adolescents and Adults in the United States." *Journal of Infectious Diseases* 196, no. 12 (2008): 1852–59.

Levant, Ronald F., Philip A. Allen, and Mei-Ching Lien. "Alexithymia in Men: How and When Do Emotional Processing Deficiencies Occur?" *Psychology of Men & Masculinity* 15, no. 3 (2014): 324–34.

Liu, Tao, and Y. Joel Wong. "The Intersection of Race and Gender: Asian American Men's Experience of Discrimination." *Psychology of Men & Masculinity* 19, no. 1 (2018): 89–101.

Lutz Zois, Catherine J., Karisa Ann Moler, and Mitchell Brown. "Mechanisms for the Association between Traditional Masculine Ideologies and Rape Myth Acceptance among College Men." *Psychology Faculty Publications*, paper 11 (2015). https://goo.gl/LxVhJ2.

Maas, Megan. "The Influence of Internet Pornography on College Students: An Empirical Analysis of Attitudes, Affect, and Sexual Behavior." *McNair Scholars Journal* 11 (2007): 137–50.

Marston, Cicely, and Ruth Lewis. "Anal Heterosex Among Young People and Implications for Health Promotion: A Qualitative Study in the UK." *BMJ Open* 4, no. 8 (2014). https://goo.gl/vwBxen.

Martellozzo, Elena, Andy Monaghan, Joanna R. Adler, et al. "'. . . I Wasn't Sure It Was Normal to Watch It . . .': A Quantitative and Qualitative Examination of the Impact of Online Pornography on the Values, Attitudes, Beliefs and Behaviours of Children and Young People." Project Report. London: Middlesex University, National Society for the Prevention of Cruelty to Children, The Children's Commissioner, 2016.

Martinez, Taylor, Jacquelyn D. Wiersma-Mosley, Kristen N. Jozkowski, et al. "'Good Guys Don't Rape': Greek and Non-Greek College Student Perpetrator Rape Myths." *Behavioral Sciences* 8, no. 60 (2018). https://doi.org/10.3390/bs8070060.

Martino, Steven, Rebecca L. Collins, Marc N. Elliott, et al. "Exposure to Degrading Versus Nondegrading Music Lyrics and Sexual Behavior Among Youth." *Pediatrics* 118, no. 2 (2006).

McKee, Alan. "Does Size Matter? Dominant Discourses about Penises in Western Culture." *Cultural Studies Review* 10, no. 2 (2004): 178.

Mellins, Claude A., Kate Walsh, Aaron L. Sarvet, et al. "Sexual Assault Incidents Among College Undergraduates: Prevalence and Factors Associated with Risk." *PLoS ONE* 12, no. 11 (2017). https://tinyurl.com/y4orl8hc.

Miller, Elizabeth, Daniel J. Tancredi, Heather L. McCauley, et al. "'Coaching Boys into Men': A Cluster-Randomized Controlled Trial of a Dating Violence Prevention Program." *Journal of Adolescent Health* 51, no. 5 (2012): 431–38.

Milburn, Michael A., Roxanne Mather, and Sheree D. Conrad. "The Effects of Viewing R-Rated Movie Scenes That Objectify Women on Perceptions of Date Rape." *Sex Roles* 43, no. 9–10 (2000), 645–64.

Morgan, Elizabeth M. "Associations Between Young Adults' Use of Sexually Explicit Materials and Their Sexual Preferences, Behaviors, and Satisfaction." *Journal of Sex Research* 48, no. 6 (2011): 520–30.

Muehlenhard, Charlene, and Sheena K. Shippee. "Men's and Women's Reports of Pretending Orgasm." *Journal of Sex Research* 47, no. 6 (2010): 552–67.

Nagoski, Emily. *Come as You Are: The Surprising New Science That Will Transform Your Sex Life.* New York: Simon & Schuster, 2015.

Norris, Jeanette, Kelly Cue Davis, William H. George, et al. "Alcohol's Direct and Indirect Effects on Men's Self-Reported Sexual Aggression Likelihood." *Journal of Studies on Alcohol* 63 (2002): 688–69.

Norris, Jeanette, William H. George, Kelly Cue Davis, et al. "Alcohol and Hypermasculinity as Determinants of Men's Empathic Responses to Violent Pornography." *Journal of Interpersonal Violence* 14 (1999): 683–700.

O'Hara, Ross E., Frederick X. Gibbons, Meg Gerrard, et al. "Greater Exposure to Sexual Content in Popular Movies Predicts Earlier Sexual Debut and Increased Sexual Risk Taking." *Psychological Science* 23, no. 9 (2012): 984–93.

Orchowski, Lindsay M., Alan Berkowitz, Jesse Boggis, et al. "Bystander Intervention Among College Men: The Role of Alcohol and Correlates of Sexual Aggression." *Journal of Interpersonal Violence* (2015): 1–23.

Orenstein, Peggy. *Girls & Sex: Navigating the Complicated New Landscape.* New York: Harper Paperbacks, 2017.

Pascoe, C.J. *Dude, You're a Fag: Masculinity and Sexuality in High School.* Berkeley: University of California Press, 2007.

Pascoe, C.J., and Jocelyn A. Hollander. "Good Guys Don't Rape: Gender, Domination and Mobilizing Rape." *Gender & Society* 30, no. 1 (2015): 67–79.

Paul, Pamela. *Pornifed: How Pornography Is Transforming Our Lives, Our Relationships, and Our Families.* New York: Times Books, 2005.

Peter, Jochen, and Patti M. Valkenberg. "Adolescents' Exposure to a Sexualized Media Environment and Their Notions of Women as Sex Objects." *Sex Roles* 56 (2007): 381–95.

———. "Adolescents' Exposure to Sexually Explicit Online Material and Recreational Attitudes Toward Sex." *Journal of Communication* 56, no. 4 (2006): 639–60.

Peralta, Robert L. "College Alcohol Use and the Embodiment of Hegemonic Masculinity Among European American Men." *Sex Roles* 56, no. 11–12 (2007): 741–56.

Pollack, William S. *Real Boys: Rescuing Our Sons from the Myths of Manhood.* New York: Random House, 1998.

Porter, Tony. "A Call to Men." TED. Lecture. December 2010.

Primack, Brian A., Erika L. Douglas, Michael J. Fine, et al. "Exposure to Sexual Lyrics and Sexual Experience Among Urban Adolescents." *American Journal of Preventive Medicine* 36, no. 4 (2009).

Ray, Rashawn. "Fraternity Life at Predominantly White Universities in the US: The Saliency of Race." *Ethnic and Racial Studies* 36, no. 2 (2013): 320–36.

Ray, Rashawn, and Jason A. Rosow. "Getting Off and Getting Intimate: How Normative Institutional Arrangements Structure Black and White Fraternity Men's Approaches Toward Women." *Men and Masculinities* 12, no. 5 (2010): 523–46.

———. "The Two Different Worlds of Black and White Fraternity Men: Visibility and Accountability as Mechanisms of Privilege." *Journal of Contemporary Ethnography* 4, no. 1 (2012): 66–94.

Reidy, Dennis E., Megan S. Early, and Kristin M. Holland. "Boys Are Victims Too? Sexual Dating Violence and Injury Among High-Risk Youth." *Preventative Medicine* 101 (2017): 28–33.

Rose, Tricia. *The Hip Hop Wars: What We Talk About When We Talk About Hip Hop and Why It Matters.* New York: Basic Books, 2008.

Rosenberger, Joshua, Michael Reece, Vanessa Schick, et al. "Sexual Behaviors and Situational Characteristics of Most Recent Male-Partnered Sexual Event among Gay and Bisexually Identified Men in the United States." *Journal of Sexual Medicine* 8, no. 11 (2011): 3040–50.

Rothman, Emily F., Elizabeth Miller, Amy Terpeluk, et al. "The Proportion of U.S. Parents Who Talk with Their Adolescent Children About Dating Abuse." *Journal of Adolescent Health* 49, no. 2 (2011): 216–18.

Rudman, Laurie A., and Eugene Borgida. "The Afterglow of Construct Accessibility: The Behavioral Consequences of Priming Men to View Women as Sexual Objects." *Journal of Experimental Social Psychology* 31, no. 6 (1995): 493–517.

Sabina, Chiara, Janis Wolak, and David Finkelhor. "The Nature and Dynamics of Internet Pornography Exposure for Youth." *CyberPsychology & Behavior* 11, no. 6 (2008).

Sanday, Peggy Reeves. "Rape-Prone Versus Rape-Free Campus Cultures." *Violence Against Women* 2, no. 2 (1996): 191–208.

Santelli, John, Stephanie A. Grilo, Tse-Hwei Choo, et al. "Does Sex Education Before College Protect Students from Sexual Assault in

College?" *PLoS One* 13, no. 11 (2018). https://doi.org/10.1371/journal.pone.0205951.

Santelli, John S., Leslie M. Kantor, Stephanie A. Grilo, et al. "Abstinence-Only-Until-Marriage: An Updated Review of U.S. Policies and Programs and Their Impact." *Journal of Adolescent Health* 61, no. 3 (2017): 273–80.

Schalet, Amy T. *Not Under My Roof: Parents, Teens, and the Culture of Sex.* Chicago: University of Chicago Press, 2011.

Scully, Diana, and Joseph Marrola. "Convicted Rapists' Vocabulary of Motive: Excuses and Justifications." *Social Problems* 31, no. 5 (1984): 530–44.

Seabrook, Rita, Monique Ward, and Soraya Giaccardi. "Why Is Fraternity Membership Associated with Sexual Assault? Exploring the Roles of Conformity to Masculine Norms, Pressure to Uphold Masculinity, and Objectification of Women." *Psychology of Men & Masculinity* 19, no. 1 (2018): 3–13.

Seabrook, Rita, and Monique Ward. "Bros Will Be Bros? The Effect of Fraternity Membership on Perceived Culpability for Sexual Assault." *Violence Against Women*, December 28, 2018. https://doi.org/10.1177/1077801218820196.

Shaffer, Catherine S., Jones Adjei, Jodi L. Viljoen, et al. "Ten-Year Trends in Physical Dating Violence Victimization Among Adolescent Boys and Girls in British Columbia, Canada." *Journal of Interpersonal Violence* (2018). https://doi.org/10.1177/0886260518788367.

Sharpley-Whiting, Tracy D. *Pimps Up, Ho's Down: Hip Hop's Hold on Young Black Women.* New York: New York University Press, 2008.

Smiler, Andrew P. *Challenging Casanova: Beyond the Stereotype of the Promiscuous Young Male.* San Francisco: Jossey-Bass, 2012.

Steinfeldt, Jesse A., Brad D. Foltz, Jessica Mungro, et al. "Masculinity Socialization in Sports: Influence of College Football Coaches." *Psychology of Men & Masculinity* 12, no. 3 (2011): 247–59.

Stemple, Lara, Andrew Flores, and Ilan H. Meyer. "Sexual Victimization Perpetrated by Women: Federal Data Reveal Surprising Prevalence." *Aggression and Violent Behavior* 34 (2017): 302–11.

Stemple, Lara. "Male Rape and Human Rights." *Hastings Law Journal* 60 (2008): 605–47.

Stermer, S. Paul, and Melissa Burkley. "SeX-Box: Exposure to Sexist Video

Games Predicts Benevolent Sexism." *Psychology of Popular Media Culture* 4, no. 1 (2015): 47–55.

Sun, Chyng, Jennifer Johnson, and Matthew Ezzell. "Pornography and the Male Sexual Script: An Analysis of Consumption and Sexual Relations." *Archives of Sexual Behavior* 45, no. 4 (2016): 983–94.

Thompson, Michael. *Raising Cain: Protecting the Inner Lives of American Boys.* New York: Ballantine, 2000.

Tolman, Deborah, Brian R. Davis, and Christin P. Bowman. "'That's Just How It Is': A Gendered Analysis of Masculinity and Femininity Ideologies in Adolescent Girls' and Boys' Heterosexual Relationships." *Journal of Adolescent Research* 31, no. 1 (2016): 3–31.

Turchik, Jessica. "Sexual Victimization Among Male College Students: Assault Severity, Sexual Functioning, and Health Risk Behaviors." *Psychology of Men & Masculinity* 11, no. 3 (2012): 243–55.

Undem, Tresa, and Ann Wang. "The State of Gender Equality for U.S. Adolescents." Warwick, RI: Plan International USA, 2018. https://goo.gl/ut3f4n.

US House of Representatives, Committee on Government Reform Minority Staff, Special Investigations Division. *The Content of Federally Funded Abstinence-Only Education Programs.* Prepared for Representative Henry A. Waxman. Washington, DC: US Government Printing Office, 2004. https://tinyurl.com/y5dheu8z.

Von Robertson, Ray, and Cassandra Chaney. "'I Know It [Racism] Still Exists Here': African American Males at a Predominantly White Institution." *Humboldt Journal of Social Relations* 1, no. 39 (2017): 260–82.

Wade, Lisa. *American Hookup: The New Culture of Sex on Campus.* New York: W. W. Norton, 2017.

Ward, L. Monique. "Understanding the Role of Entertainment Media in the Sexual Socialization of American Youth: A Review of Empirical Research." *Developmental Review* 23 (2003): 347–88.

Ward, L. Monique, and Kimberly Friedman. "Using TV as a Guide: Associations Between Television Viewing and Adolescents' Sexual Attitudes and Behavior." *Journal of Research on Adolescence* 16 (2006): 133–56.

Ward, L. Monique, Edwina Hansbrough, and Eboni Walker. "Contributions of Music Video Exposure to Black Adolescents' Gender and Sexual Schemas." *Journal of Adolescent Research* 20 (2005): 143–66.

Way, Niobe. *Deep Secrets: Boys' Friendships and the Crisis of Connection.* Cambridge, MA: Harvard University Press, 2013.

Weinberg, M. Katherine, Edward Tronick, Jeffrey Cohn, et al. "Gender Differences in Emotional Expressivity and Self-Regulation During Early Infancy." *Developmental Psychology* 35, no. 1 (1999): 175–88.

Weissbourd, Richard, Trisha Ross Anderson, Alison Cashin, et al. "The Talk: How Adults Can Promote Young People's Healthy Relationships and Prevent Misogyny and Sexual Harassment." Cambridge, MA: Making Caring Common Project Harvard Graduate School of Education, 2017.

West, Cornel. *Race Matters*, anniversary edition. Boston: Beacon Press, 2017.

Widen, Sherri C., and James A. Russell. "Gender and Preschoolers' Perception of Emotion." *Merrill-Palmer Quarterly* 48, no. 3 (2002): 248–61.

Widman, Laura, Reina Evans, Hannah Javidi, et al. "Assessment of Parent-Based Interventions for Adolescent Health: A Systematic Review and Meta-Analysis." *JAMA Pediatrics* 173 (2019): 866–77.

Wiseman, Rosalind. *Masterminds and Wingmen: Helping Our Boys Cope with Schoolyard Power, Locker-Room Tests, Girlfriends, and the New Rules of Boy World.* New York: Harmony Books, 2013.

Wolak, Janis, Kimberly Mitchell, and David Finkelhor. "Unwanted and Wanted Exposure to Online Pornography in a National Sample of Youth Internet Users." *Pediatrics* 119, no. 2 (2007): 247–57.

Wong, Y. Joel, Moon-Ho Ringo Ho, Shu-Yi Wang, et al. "Meta-Analyses of the Relationship Between Conformity to Masculine Norms and Mental Health-Related Outcomes." *Journal of Counseling Psychology* 64, no. 1 (2017): 80–93.

Wong, Y. Joel, Angela J. Horn, and Shitao Chen. "Perceived Masculinity: The Potential Influence of Race, Racial Essentialist Beliefs, and Stereotypes." *Psychology of Men & Masculinity* 14, no. 4 (2013): 452–64.

Wong, Y. Joel, Jesse Owen, Kimberly K. Tran, et al. "Asian American Male College Students' Perceptions of People's Stereotypes About Asian American Men." *Psychology of Men & Masculinity* 23, no. 1 (2012): 75–88.

Wong, Y. Joel, Jesse A. Steinfeldt, Julie R. LaFollette, et al. "Men's Tears: Football Players' Evaluations of Crying Behavior." *Psychology of Men & Masculinity* 12, no. 4 (2011): 297–310.

Wright, Paul J. "Show Me the Data! Empirical Support for the 'Centerfold Syndrome.'" *Psychology of Men & Masculinity* 13, no. 2 (2012): 180–98.

———. "A Three-Wave Longitudinal Analysis of Preexisting Beliefs, Exposure to Pornography, and Attitude Change." *Communication Reports* 26, no. 1 (2013): 13–25.

Wright, Paul J., Ana J. Bridges, Chyng F. Sun, et al. "Personal Pornography Viewing and Sexual Satisfaction: A Quadratic Analysis." *Journal of Sex and Marital Therapy* 44, no. 3 (2018): 308–15.

Wright, Paul J., Chyng F. Sun, Nicola J. Steffen, et al. "Associative Pathways Between Pornography Consumption and Reduced Sexual Satisfaction." *Sexual and Relationship Therapy* (2017). https://doi.org/10.1080/14681994.2017.1323076.

———. "Pornography, Alcohol and Male Sexual Dominance." *Communication Monographs* 82, no. 2 (2015): 252–70.

Wright, Paul J., and Michelle Funk. "Pornography Consumption and Opposition to Affirmative Action for Women: A Prospective Study." *Psychology of Women Quarterly* 38, no. 2 (2013): 208–21.

Wright, Paul J., Neil M. Malamuth, and Ed Donnerstein. "Research on Sex in the Media: What Do We Know About Effects on Children and Adolescents?" In *Handbook of Children and the Media*, 2nd ed., edited by Dorothy G. Singer and Jerome L. Singer. Los Angeles: Sage Publications, 2012.

Wright, Paul J., and Ashley K. Randall. "Pornography Consumption, Education, and Support for Same-Sex Marriage Among Adult U.S. Males." *Communication Research* 41, no. 5 (2013): 665–89.

Wright, Paul J., Nicola J. Steffen, and Chyng F. Sun. "Is the Relationship Between Pornography Consumption Frequency and Lower Sexual Satisfaction Curvilinear? Results from England and Germany." *Journal of Sex Research* 56 (2019): 9–15.

Wright, Paul J., and Robert S. Tokunaga. "Activating the Centerfold Syndrome: Recency of Exposure, Sexual Explicitness, Past Exposure to Objectifying Media." *Communications Research* 20, no. 10 (2013): 1–34.

———. "Men's Objectifying Media Consumption, Objectification of Women, and Attitudes Supportive of Violence Against Women." *Archives of Sexual Behavior* 45, no. 4 (2016): 955–64.

Wright, Paul J., Robert Tokunaga, Ashley Kraus, et al. "Pornography

Consumption and Satisfaction: A Meta-Analysis." *Human Communication Research* 43, no. 3 (2017): 314–43.

Yasso, Tara, William A. Smith, Miguel Ceja, et al. "Critical Race Theory, Racial Microaggressions, and Campus Racial Climate for Latina/o Undergraduates." *Harvard Educational Review* 79, no. 4 (2009): 659–91.

Yao, Mike Z., Chad Mahood, and Daniel Linz. "Sexual Priming, Gender Stereotyping, and Likelihood to Sexually Harass: Examining the Cognitive Effects of Playing a Sexually-Explicit Video Game." *Sex Roles* 62, no. 102 (2010): 77–88.

Zaloom, Shafia. *Sex, Teens, and Everything in Between: The New and Necessary Conversations Today's Teenagers Need to Have about Consent, Sexual Harassment, Healthy Relationships, Love, and More.* Naperville, IL: Sourcebooks, 2019.

Zhang, Yuanyuan, Laura Miller, and Kristen Harrison. "The Relationship Between Exposure to Sexual Music Videos and Young Adults' Sexual Attitudes." *Journal of Broadcasting & Electronic Media* 52, no. 3 (2008): 368–86.

Index

AAU Campus Climate Survey, 169
abstinence-only education, 199, 220
abuse
  physical, 101
  sexual, 191–93
Access Hollywood bus, Trump comments on, 27–28
accountability
  African American perspective on, 162–63
  for sexual assault, 197–99, 211–18
Adam, on masculinity, 25–26
Adult Swim, 61, 66
affirmative consent, 169–71, 222–23
African Americans
  as "cool," 140–42
  fraternities of, 162–63
  higher education experiences of, 135–38, 158–63
  hookups of, 142–44, 155–58, 162–64
  masculinity of, 140–42
  sexism against, 67
  sexual assault and, 143–44, 152–56
  sexual racism against, 149–51
  social integration of, 138–40
aggression
  boys of color and, 143–44
  as masculine trait, 10–13, 22
  parent talks on, 234–37
  porn and, 49, 57, 59–61
  sexual assault and, 177–78, 187–88
  in talk about sex, 28–29
Aidan, on experience as boy of color, 140–41, 155–56
AIDS, 106, 108, 226
Alan, on unwanted sex, 191
alcohol
  African American consumption of, 162
  consent misreading and, 172–73
  in hookup culture, 78–79, 145–47
  sexual assault and, 78–79, 172–73
Amherst, sexism scandal at, 30
anal sex
  among gay guys, 120–22, 226
  coercion and, 171
  gay porn portrayal of, 109
  as "not sex," 226
  porn influence on, 48, 56–57
Andrew, on hookups, 78, 88
Angels in America (Kushner), 106

anger
as masculine emotion, 12, 15–16, 22
after sexual assault, 187–88
Ansari, Aziz, 175
Anwen, on restorative justice for sexual assault, 199–211, 214–18
Armstrong, Elizabeth, 136–37
arousal non-concordance, 57–59
Asian Americans
hookups of, 144–49, 164
sexual racism against, 148–52
Askmen.com, on acceptable crying, 18–19
athletes
assault perpetration by, 29, 235
masculinity of, 7–9, 19–22
sexism among, 29–31, 112–13, 118–19, 234–37
tears shed by, 18–19
athleticism, 10–11
athletics. *See* sports
Audrie and Daisy, 33

barebacking, 109
bathroom bills, 114
Bedera, Nicole, 170–72, 176
Berkeley, sexism scandal in, 31
Big Sick, The, 148
Billions, 64
binge-drinking, 78–79, 145–47
bisexualism, 106
bitch
emotional expression as quality of, 12
use of term, 22, 37
black guys. *See* African Americans
body discomfort, 114–16
body image, 49–51
boy code, 13, 24
boys
expectations of, 1–2
girl views of, 11
interview experiences with, 3–6
porn portrayal of, 43
traits of ideal, 9–12
boys of color
fraternities of, 162–63
higher education experiences of, 135–38, 158–63
hookups of, 142–50, 155–58, 162–64
masculinity of, 140–42, 149–52
sexual assault and, 143–44, 152–56
sexual racism against, 148–52
social integration of, 138–40
Brandon, on consent and sexual assault, 176–77
breakups, 101–2
bro culture, 7–9, 20–22, 234–37
Brown, Brené, 16
Brown, Chris, 141
Buttigieg, Pete, 107

Caleb, on hookups, 89–92
Californication, 65
casual sex. *See* hookups
Centers for Disease Control, 220
Chance the Rapper, 155
Chen, Nicole, 149
Chu, Judy Y., 16
Cirioni, Frank, 211, 213–14, 216
cis, concept of, 133
climax. *See* orgasm
CM Punk, 27

coaches, 20–21, 236–37
Cole
on masculinity, 7–9, 14–15, 21–22, 34–37
on porn, 45
Coleman, Daisy, 33
college experience, views on, 136–37
Columbia University
sexism scandal at, 29
Sexual Health Initiative to Foster Transformation (SHIFT), 185, 193
coming out
average age of, 107
as gay, 109–10
as transgender, 114
communication. *See also* talk about sex
in hookups, 75, 89–92
compulsion, porn as, 52
compulsory carelessness, of hookups, 78
confidence, 166
connection. *See* emotional connection
consent
affirmative, 169–71, 222–23
as bare minimum, 100
boys as grantors of, 183–96, 223–24
debates on, 168–69
encounters at edges of, 175–76
among gay guys, 120–22
in hookups, 94–95, 100, 165–68, 176–80
parent talks on, 180–82, 222–24
sex education coverage of, 220
signals mistaken as, 171–73
transgender perspective on, 117–18
uncertainty about, 165–68, 176–78
control, 154–55
Cook, Tim, 107
Cosby, Bill, 155
Cox, Laverne, 114
cross-racial dating, 148–49
crying, 12, 18–19

Daniel, on porn, 55–57
dating apps, 122–27
Denison, Charis, 34, 236
Devon, on transgender experience, 112–19, 128–33
DeVos, Betsy, 199
dick pic, 123
"dick school," 34
parent talks on, 234–37
Dietrich, Savannah, 33
Dill-Shackleford, Karen, 46–47
DMs, after hookups, 92
dominance
as ideal guy trait, 10–11
porn influence on, 49
Dylan, on unwanted sex, 183–84, 186–88, 194–96

education. *See* sex education
Ellie, on hookups, 89–92
Elliot, on gay experience, 123–24
Ely, on masculinity, 12
Eminem, 194
Emmett, on experience as boy of color, 136–37, 142–43, 159–64
emotional connection, 15, 24
in hookups, 93–104, 127–28
in relationships, 99–104

emotional expression
after hookups, 91
ideal guy, 9–11
infant boy, 15–16
masculinity constraints on, 11–19, 31–37
parent expectations of, 15–16, 228–32
after sexual assault, 187–88, 193–94
erections, porn influence on, 51, 56–57, 69–70
Eric, on masculinity, 20
Ethan, on masculinity, 20–21
ethical sex, 224–25
expulsions, of boys of color, 152–55

fag, use of term, 22–27, 36
fake orgasm, 189
Family Guy, 61
fantasy slut league, Piedmont, 30
fathers. *See also* parents
consent guidance from, 180–82
emotional expression expectations of, 15–16
masculinity messages from, 11–12, 14–16, 228–32
Federer, Roger, 19
feelings. *See also* emotional expression
in hookups, 93–104, 127–28
in relationships, 99–104
female masculinity, 131
female-to-male transition, 116
femininity, masculinity as opposite of, 13–14
feminism, 13
fetish videos, 52, 54–55
Fifty Shades of Grey, 58
fitness gurus, 40
Ford, Christine Blasey, 231
Ford, Jessie, 188–90, 193
fraternities
assault by members of, 145, 174, 185, 198, 214, 235
black, 162–63
hookups in, 77, 147
parent talks on, 234–37
racism in, 159–61
sexism in, 21, 30, 161, 162, 178–79
"Freaky Friday" (Lil Dicky), 141

gangsta-pimp-ho trinity, 67
gay, use of term, 23–27, 120
gay guys
acceptance of, 111–12
broadened view of sex for, 226
challenges faced by, 108
coming out experience for, 109–10
consent among, 120–22
current attitudes toward, 106–8
dating app use by, 122–27
hookups among, 105, 120–28
masculinity of, 25–27
porn use by, 108–9
sex talk with, 125, 226
sexual assault of, 126–27
support of, 23, 110–11
terminology used by, 120
gendered racism, 150
gender identity. *See also* transgender guys
current attitudes toward, 106–8
question of, 133
gender socialization, 15–16, 214, 219
genitalia
first glimpse of, 44–45
porn influence on, 51

genital response, 57–59
Gen Z, LGBTQ+, 107
Get Out, 148
girlfriends
  boys' descriptions of, 100
  emotional expression with, 16–17
girlieness, 13–14
girls
  emotional expression as acting like, 12
  media sexualization of, 62–64
  porn portrayal of, 43, 55, 61–64
  satisfaction of, 80–81, 88
  views about, 11
Glamour, 231
Golden Dick Syndrome, 93
gonorrhea, 108
"good guys," sexual assault by, 173–80
good sex, definition of, 224
gossip, 87–88
GQ
  on acceptable crying, 18
  #MeToo survey, 231
Greek life. *See* fraternities
Green Dot, 207–8
Grindr, 123–27

Halberstam, Jack, 131–33
Hamilton, Laura T., 136–37
hand holding, 99
Harmon, Dan, 66, 231
Harvard scouting report scandal, 29, 231
Harvey, Steve, 150
Herman, Judith Lewis, 212
higher education. *See also* fraternities
  boys of color in, 135–38, 158–63
  sexual assault in, 168–69
hilarious, 47, 61
  unwanted sex as, 193
  use of term, 32–34
hip-hop, 66–68
Hirsch, Jennifer, 185, 193
HIV, 106, 108, 226
Hollander, Jocelyn, 174
homophobia, 21
hookups, 73–74
  of African Americans, 142–44, 155–58, 162–64
  alcohol in, 78–79, 145–47
  of Asian Americans, 144–49, 164
  awkwardness during, 78
  communication in, 75, 89–92
  confidence in, 166
  consent in, 94–95, 100, 165–68, 176–80
  definition of, 76
  gay, 105, 120–28
  intercourse in, 76–77
  of Latinx guys, 150
  lessons taken from, 102
  masculinity in, 79–80, 83, 128–29
  objectification of women in, 79–80, 100
  orgasms in, 80–81, 88
  reputation and, 82–88
  respect and love in, 93–104
  sexual conquest in, 93–94
  sexual satisfaction in, 80–81, 88, 90, 100
  texts after, 89–92
  transgender guys in, 128–30
  wins and losses in, 84–86
hormone therapy, 116
Huang, Eddie, 151
Hughey, Matthew, 161

humor, sexist, 31–34
Hutch, on transgender experience, 117–18, 131

I Am Jazz, 114
ideal guy, 9–12
impact statement, 213–14, 231
infant boys, 15–16
intercourse
  in hookups, 76–77
  rates of, 50, 76
internet, porn availability on, 41–45, 59
intimacy
  African American perspective on, 162, 164
  in hookups, 83, 93–104, 127–28
  parent talks on, 180–82, 224–25, 228–32
  in relationships, 99–104
  after sexual assault, 193–94
It Gets Better Project, 121

Jackson, Michael, 192
Jay-Z, 67
Jenner, Caitlyn, 114
Jeong, Ken, 151
jock culture, 7–9, 19–22, 234–37
jokes
  sexist, 31–34
  unwanted sex as, 193
Josh, on masculinity, 19–20

Karp, David, 212–13
Kavanaugh, Brett, 213, 231
Kevin, on porn, 50
Kimmel, Michael, 141
#KindrGrindr, 123
King, Rodney, 138
Kingsman, 56
Kushner, Tony, 106

lad magazines, 63–64
Latinx guys, 150
LeeAnn, on hookups, 89–92
Leo, on unwanted sex, 188–91
Lepp, Stephanie, 199
lesbian porn, 108
LGBTQ+. *See also* gay guys; transgender guys
  acceptance of, 111–12
  challenges faced by, 108
  current attitudes toward, 106–8
  numbers of, 107
  parental discussion of, 109–11, 114, 116, 226
  porn use by, 108–9
  support of, 23, 110–11
Liam, on consent and sexual assault, 165–69, 174, 180–82, 194
Lil Dicky, 141
Lipsyte, Robert, 21
locker room banter, 8–9, 20–22
  sexual conquest in, 27–29
  transgender perspective on, 112–13, 118–19
losses, in hookups, 84–86
Louis C. K., 175
love
  in hookups, 93–104
  parent talks on, 224–25
  in relationships, 99–104
Luke, on masculinity, 26–27

magazines, sexism in, 63–64
Making Caring Common, 181–82, 225
man box, 13, 23, 228

Marshall, Kendall, 27
masculinity
  aggression and, 10–13, 22
  of boys of color, 140–42, 149–52
  bro culture and, 7–9, 19–22, 234–37
  emotional expression constraints of, 11–19, 31–37
  fag use in, 22–27, 36
  of gay guys, 25–27
  in hookup culture, 79–80, 83, 128–29
  ideal traits of, 9–12
  as opposite of femininity, 13–14
  parent talks on, 11–12, 14–16, 228–32
  power of traditional conception of, 13
  in reactions to unwanted sex, 190–94
  sexism and, 27–34
  sexual conquest as, 10–11, 13, 27–34
  sexual performance and, 88
  silence in, 34–37, 236–37
  toxic, 2, 13, 19, 27, 129, 228–32
  transgender perspective on, 117–19, 128–32
Mason
  first impression of, 39–41
  on porn, 52–55, 65, 68–71
masturbation, 42
  parent talks on, 232–33
  porn tie to, 44–45
  in range of contexts, 58–59, 232–33
  restraint from, 52
Mateo, on masculinity, 25, 28
Mathis, Dara, 68
Mauricio, on experience as boy of color, 150
Mean Girls, 219
media
  rape in, 64
  sexism in, 61–66
Mellins, Claude Ann, 185–86
#MeToo, 67, 231, 237
millennials
  internet use by, 41
  LGBTQ+, 107
  rates of intercourse among, 50, 76
Miller, Chanel, 231
MindGeek, 43
Mindy Project, The, 48
misogyny
  in hip-hop, 67–68
  parental talk about, 227–28
  parent talks on, 180–82
  racism coexistence with, 161
  transgender perspective on, 112–13, 117–19
Mitchell, on porn, 45, 51
Modern Family, 65
mothers. *See also* parents
  of boys committing assault, 198–99
  emotional expression with, 15–18
  talking about intimacy, 230

Nagoski, Emily, 57–59, 232
Nas, 67
Nassar, Larry, 192
Nate, on hookups, 81–88, 101–4
Netherlands
  attitudes toward teen sexuality in, 97–98
  sex education in, 233–34
Nick, on consent and sexual assault, 178–79

Noah
on masculinity, 15, 20
on talking with parents, 229–30
Noah, Trevor, 27
"no fap" movement, 52
#nohomo, 24
non-concordance, 57–59
nonconsensual sex. *See* rape

Obama administration, 168
objectification of women. *See* sexism
Online College Social Life Survey, 77, 80
oral sex, 51, 76, 121, 143, 172–75
orgasm
fake, 189
in hookups, 80–81, 88
porn influence on, 56–57

parents. *See also* talk about sex
of boys committing assault, 198–99
coming out to, 109–10, 114
consent guidance from, 180–82, 222–24
emotional expression expectations of, 15–16
masculinity messages from, 11–12, 14–16, 228–32
porn use and, 43–44, 52–54
Pascoe, C.J., 23–24, 174
Paul, Bryant, 60–61
Paying for the Party (Armstrong and Hamilton), 137
Peele, Jordan, 148
penis, porn and, 51
performance
in hookups, 87–88
porn influence on, 51–52
physical abuse, 101
physical appearance, 9–11
Piedmont fantasy slut league, 30
Pi Kappa Phi, 161
pleasure
for both partners, 125, 143, 176
in hookups, 80–81, 88, 90, 100
male prioritization of, 175–76
porn reduction of, 49–50
in relationships, 99–101
Pollack, William, 13
porn
access to, 41–45, 59
aggression and, 49, 57, 59–61
as compulsion, 52
definition of, 42
desensitization by, 59
first exposure to, 44–45, 48–49
gay, 108–9
genital response and, 57–59
masturbation tie to, 44–45
parent talks on, 232–33
prevalence of, 43–44
rape and, 49, 62
reality compared with, 43, 46–52, 68–71, 109
rejection of, 45–46
sexism and, 49, 52–55, 61
transgender perspective on, 129–30
unconscious influences of, 46–52, 56–57, 68–71
Pornhub, 42–43, 45, 109
power, sexual conquest as, 28–29
precocious sexuality, 47–48
PrEP, 226
priming theory, 61

protection, 226
pussy
  emotional expression as quality of, 14
  use of term, 22, 37

queer, use of term, 107, 120

race. *See* boys of color
racism
  in higher education, 159–62
  sexual, 148–52
rape
  affirmative consent standard for, 169–71, 222–23
  alcohol consumption and, 78–79
  boys of color and, 154–56
  by "good guys," 173–80
  "hilarious" as defense of, 32–33
  media portrayal of, 64
  misconceptions of, 169
  non-concordance in, 58
  porn and, 49, 62
  racism coexistence with, 161
  rationalizations of, 171
  report of, 198
  transgender perspective on, 117–18
reality, porn compared with, 43, 46–52, 68–71, 109
relationships
  African American perspective of, 162, 164
  hookups and, 77
  love, respect, and sex in, 99–104
  parent talks on, 224–25
  after sexual assault, 193–94
  vulnerability in, 16
reputation, 82–88
respect
  African American perspective on, 162
  in hookups, 93–104
  parent talks on, 180–82
  in relationships, 99–104
responsibility, 197–99, 211–18
restorative justice (RJ) for sexual assault, 197–218
Reza
  on consent and sexual assault, 177–78
  on porn, 50
Rick and Morty, 66
RJ. See restorative justice
R. Kelly, 155
Rob, on masculinity, 12, 16–19, 23–24
Roof, Dylann, 154
Rose, Tricia, 67
"Rule #34," 42

Sameer, on restorative justice for sexual assault, 199–211, 214–18
same-sex marriage, 107
Savage, Dan, 121, 124
Schalet, Amy, 98, 226, 233
scouting report scandal, Harvard, 29, 231
sex. *See also* hookups
  curiosity about, 42
  definition of, 121, 225–26
  disappointment in, 77–78
  porn influence on, 47–51, 56–57, 68–71
  transgender experience of, 114–15, 131–33

sex education
abstinence-only, 199, 220
curricula for, 108
insufficiency of, 219–21
in Netherlands, 233–34
porn use for, 43
sexism
African Americans and, 162
among athletes, 29–31, 112–13, 118–19, 234–37
in hip-hop, 66–68
hookup culture and, 79–80, 100
masculinity and, 27–34
in media, 61–66
parent talks on, 180–82, 227–28
porn and, 49, 52–55, 61
transgender perspective on, 112–13, 117–19, 128–29
in video games, 66
sex talk. *See* talk about sex
sexual abuse, 191–93
sexual assault
affirmative consent standard for, 169–71, 222–23
alcohol and, 78–79, 172–73
boys as victims of, 183–96, 223–24
boys of color and, 143–44, 152–56
boys' reactions to, 187–89, 187–96
"Dear Colleague" letter on, 168
among gay guys, 126–27
by "good guys," 173–80
"hilarious" as defense of, 32–34
media portrayal of, 64
parent talks on, 180–82, 228–32, 234–37
pervasiveness of, 1–2
porn influence on, 62
racism coexistence with, 161
rates of, 168–69
rationalizations of, 171
report of, 198
restorative justice for, 197–218
transgender perspective on, 117–18
uncertainty about consent in, 165–68, 176–78
sexual conquest
in hookups, 93–94
in jock culture, 21
masculinity and, 10–11, 13, 27–34
sexual harassment
"hilarious" as defense of, 32–33
parent talks on, 180–82, 228–32
traditional conception of masculinity and, 13
Sexual Health Initiative to Foster Transformation (SHIFT), 185–86, 193
sexuality
acceptance of, 111–12
current attitudes toward, 106–8
question of, 133
sexually transmitted disease, 106, 108, 226
sexual performance. *See* performance
sexual pleasure. *See* pleasure
sexual racism, 148–52
sexual response, porn and, 57–59
Shavershian, Aziz, 40
SHIFT. *See* Sexual Health Initiative to Foster Transformation
Sigma Alpha Epsilon, 160–61
silence, masculinity rules for, 34–37, 236–37
Silicon Valley, 64

Silicon Valley high schools, sexism scandal in, 30–31
sleepovers, 233–34
smartphone, porn access on, 41–43
Smiler, Andrew, 225, 227
Snapchats, after hookups, 92
sober monitors, 144
social control, 154–55
social integration, 138–40
socialization, gender, 15–16, 214, 219
Spacey, Kevin, 192
Spencer, on experience as boy of color, 144–52, 164
sports
  jock culture of, 7–9, 19–22, 234–37
  tears shed in, 18–19
straight, concept of, 133
submissiveness, 48–49
Surviving R. Kelly, 155
Swarthmore sexism scandal, 30
swipe apps, 122–27
synthol, 40
syphilis, 108

taboo
  porn portrayal of, 60
  response to, 58
talk about sex, 2, 125
  broadened views, 225–26
  bro culture, 234–37
  consent, 180–82, 222–24
  ethical sex, 224–25
  masculinity, 228–32
  need for, 219–22
  porn, 232–33
  sexism, 227–28
  sleepovers, 233–34
  "the talk" compared with, 222
  teenager desire for, 221–22
tears, 18–19
Tebow, Tim, 19
texting, after hookups, 89–92
Theo, on consent and sexual assault, 179–80
Theta Tau, 161
Thompson, Michael, 34
THOTs of Berkeley High, 31
Till, Emmett, 154
Title IX, 168
toxic masculinity, 2, 13, 19, 27, 129, 228–32
trans, concept of, 133
transgender guys
  body discomfort and, 114–16
  coming out experience for, 114
  experience of sex as, 114–15, 131–33
  hookups of, 128–30
  masculinity from perspective of, 117–19, 128–32
  sexism from perspective of, 112–13, 117–19, 128–29
  support for, 114
Transparent, 114
Trump, Donald, 27–28, 191
Turner, Brock, 75, 174–75, 231
Tyndall, George, 192

unwanted sex. *See also* rape; sexual assault
  boys as victims of, 183–96, 223–24
  boys' reactions to, 187–96

vagina, first glimpse of, 44–45
Vegan Gains, 40
video games, 66

violence
boys as victims of, 184–86
in talk about sex, 28–29
traditional conception of masculinity and, 13
virginity, loss of, 70, 120, 146–47, 183–84, 188, 195
vulnerability
in hookups, 91, 97–98, 102, 128
masculinity constraints on, 12, 14–15, 19
in sex, 71

Wade, Lisa, 77–78, 80, 90–91, 102
wealth, as ideal guy trait, 10–11
Weinstein, Harvey, 169, 231
Weissbourd, Richard, 181–82, 225
West, Kanye, 67–68, 93
wins, in hookups, 84–85
Wolf of Wall Street, The, 64
Wong, Y. Joel, 149–50
Wright, Paul, 48–49
Wyatt
on affirmative consent, 170
on hookups, 93–97, 99–104

Xavier, on experience as boy of color, 136–42, 152–61

Yang, Frank, 40

Zachary, on sexual consent, 170
Zaloom, Shafia, 224
Zane, on gay experience, 105, 109–12, 119–28